Manischa Partowi, Annette Urban, Manuel van der Veen (eds.)
Virtual Reality Exhibited

Virtual Lifeworlds | 9

Editorial

Virtual Lifeworlds is a book series emerging from the DFG Collaborative Research Center 1567. It brings together interdisciplinary perspectives on how virtuality has become normalized. By tracing knowledge production, practices, negotiation processes, and network dynamics, the series shows how virtual lifeworlds have taken shape and how diverse forms of virtuality have become drivers of social and cultural change. It examines the functions and consequences of the virtual for the constitution of subjects, lifeworlds, and aesthetic practices; for social organization and processes; and for the scientific disciplines themselves.

The Collaborative Research Center unites education, history, art history, linguistics, literary studies, media studies, and the social sciences. Across these fields, research converges on the media and technological conditions under which virtual worlds are produced and perceived – worlds that can be narrated, calculated, experienced immersively, modeled, or imagined. The series foregrounds the concept of virtuality, the uses of virtual spaces, modes of interaction with them, and forms of participation in them. It maps the infrastructures and imaginaries that sustain virtual environments, reflects on their historical trajectories, and illuminates the ways virtuality reshapes everyday life, institutions, and knowledge.

The series is edited by Stefan Rieger, Florian Sprenger and Anna Tuschling. They represent the CRC 1567 *Virtual Lifeworlds.*

Manischa Partowi is a researcher interested in affective strategies in contemporary art. In her PhD project, she investigates the aesthetic constellation of virtual reality-based art by focusing on the phenomena of intimacy and violence. Trained in art history, philosophy and cultural studies, she is affiliated with the Collaborative Research Center 1567 *Virtual Lifeworlds* at Ruhr-Universität Bochum.

Annette Urban (Prof. Dr.) is a professor of modern and contemporary art history with a focus on new media at the Institute of Art History at Ruhr-Universität Bochum. She is principal investigator of the sub-project *Virtual Art* within the Collaborative Research Center 1567 *Virtual Lifeworlds* and a member of the Research Training Group 2132 *The Documentary. Excess and Privation.* Her research also focuses on site-relatedness in media art and digital (im)material labour in contemporary art.

Manuel van der Veen (Dr.) is a research assistant in the sub-project *Virtual Art* within the Collaborative Research Center 1567 *Virtual Lifeworlds* at Ruhr-Universität Bochum. Previously, in 2022, he completed his doctorate on the art history of augmented reality. His further research interests include the philosophy of technology and space in relation to contemporary digital art, intersections between new media and painting as well as the tradition of three-dimensional image carriers.

Manischa Partowi, Annette Urban, Manuel van der Veen (eds.)

Virtual Reality Exhibited

Interfacing Art, Games and Everyday Life

[transcript]

Funded by the Deutsche Forschungsgemeinschaft (DFG, German Research Foundation) – Collaborative Research Centre 1567 *Virtual Lifeworlds* (470106373).

Bibliographic information published by the Deutsche Nationalbibliothek
The Deutsche Nationalbibliothek lists this publication in the Deutsche Nationalbibliografie; detailed bibliographic data are available online at https://dnb.dnb.de

transcript Verlag | Hermannstraße 26 | D-33602 Bielefeld | live@transcript-verlag.de

Cover design: Maria Arndt
Cover illustration: © Banz & Bowinkel, wire mesh illustration of the installation draft for *Palo Alto* (2018) at the NRW-Forum (Museum Kunstpalast), Düsseldorf. Courtesy of the artists.
Printing: Druckhaus Bechstein GmbH, Wetzlar
Translation of the Introduction: Marie-Alix Isdahl
https://doi.org/10.14361/9783839407554
Print-ISBN: 978-3-8376-7772-0 | PDF-ISBN: 978-3-8394-0755-4
ISSN of series: 2943-6915 | eISSN of series: 2943-6923

Printed on permanent acid-free text paper.

Contents

GAMES/EVERYDAY LIFE
Concepts of VR

EVERYDAY LIFE/ART
Modes of Reception in VR

Appendix

Acknowledgements

The contributions in this volume are based on the workshop *Virtual Reality. Interfacing Art, Games and Everyday Life*, which was organized as part of the CRC *Virtual Lifeworlds*. It took place on 23–24 January 2025 at Ruhr University Bochum. Additional authors were recruited for the publication. We would like to thank all speakers and contributors for their inspiring research and dedicated collaboration. We would also like to take this opportunity to thank our student assistants Jasmin Kaya and Lennart Schmidt for their editorial contribution to this anthology. We are grateful to Ina Bolinski, Sylvia Kokot, and Felix Rissel from the CRC *Virtual Lifeworlds* for their administrative support during the publication process.

The editors, Bochum, October 2025

The Renewed Artistic Interest in VR

An Introduction to Research on Virtual Art

Manischa Partowi, Annette Urban, Manuel van der Veen

Virtual reality (VR) is booming in art, and at the same time it is still seeking its place. This twofold insight, which has thoroughly shaped our art historical research,[1] becomes evident with a closer look at the adjacent fields of art production and reception, art mediation and criticism, as well as academic discourse. On the one hand, dedicated art prizes, residencies, and related monographs have featured VR[2] as a promising new genre of digital art that deserves active support. Some exhibitions and conferences have initiated debates addressing the consequences that VR, as an emerging medium for artists, could have on contemporary art in general and on the role of museums today.[3] On the other hand, major resistance and obstacles to the medium's acceptance still exist among curators, art enthusiasts, and the discipline of art history. This resistance is caused by obvious access barriers, technical complexities and imperfections. It also stems from a mismatch with well-established aesthetic

1 The authors of this introduction are collaborating on the subproject C03 *Virtual Art: World, Object, and Body References in Contemporary VR Experiences* as part of the CRC 1567 *Virtual Lifeworlds at Ruhr University Bochum* (2022-) and have organized the international workshop *Virtual Reality. Interfacing Art, Games and Everyday Life* (23.-24.01.2025). Its proceedings are published in this volume.

2 Tina Sauerländer (2023): Unleashed Utopias. Artistic Speculations about Today and Tomorrow in the Metaverse, Ex. Cat., VR Art Prize Berlin, online: https://vrkunst.dkb-stiftung.de/export/sites/vrkunst/en/documents/UU_katalog_EN_final.pdf (last access: 15.09.2025); Paul Luckraft/Neilson Elizabeth (eds.) (2021): Artists in Virtual Reality, London: Zabludowicz Collection.

3 See for example Lívia Nolasco-Rózsás/Marianne Schädler (eds.) (2023): Beyond Matter, Within Space. Curatorial and Art Mediation Techniques on the Verge of Virtual Reality, Berlin: Hatje Cantz Verlag; 37. Deutscher Kongress für Kunstgeschichte, 2024, Section: »Bildräumlichkeit/Raumbildlichkeit – Paradigmatische Wechselbeziehungen und Übergänge ausgehend von VR« chaired by Annette Urban and Stephan Günzel.

premises, including the fundamental category of the artwork which is traditionally understood to require a certain distance in order to be properly encountered.[4] In addition, social unease and unfulfilled expectations[5] are often prevalent, caused not least by the cyclical resurgence of VR promises and their failures.[6] Nevertheless, the broader art public can nowadays hardly avoid coming into contact with VR-based artworks, which are no longer limited to specialized exhibition venues or festivals. Rather, visitors to contemporary art exhibitions are becoming increasingly accustomed to wearing a head-mounted display (HMD) to access artistic VR experiences, even if this mode of experience breaks sharply with the conventions of museum spaces and visitors are known to change their habits only slowly.

This progressive but not frictionless emergence of VR in art museums points to a changing terrain, which can be described as a fundamental normalization of virtual technologies. As these technologies are now embedded in almost every sector of society, they can be understood as a kind of Verlebensweltlichung [becoming lifeworld],[7] as the Bochum Collaborative Research Centre (CRC) *Virtual Lifeworlds* proposes. In this respect, the conditions for artists working with VR today have significantly changed in comparison to the first wave of VR art in the 1990s.[8] Some of the early canonical VR artworks such as *Home of the Brain* (1992) by Monika Fleischmann and Wolfgang Strauss had already entered prominent art institutions such as Mies van der Rohe's Neue Nationalgalerie in Berlin. But most of these works were presented in highly specialized contexts, often in sites of their production, such as newly established media art academies, centers, and accompanying festivals. As the title of Fleischmann and Strauss's work suggests, these early works tended

4 This reservation continues to apply to immersive art in general, and even more to the image spaces in VR, which envelop viewers as environments centred on the viewer. Cf. among others Julia Reich (2023): »What to Do in and with Images? Forms of Hands-On Activity in Augmented and Virtual Reality«, in: Maria Bremer/Markus Heinzelmann (eds), Diving into Art, Ex. Cat., Wien: Verlag für Moderne Kunst, pp. 81–89.

5 See for example the contributions by Manischa Partowi and Federica Cavaletti/Pietro Conte/Andrea Pinotti in this volume.

6 See Marcus Carter/Ben Egliston (2024): Fantasies of Virtual Reality. Untangling Fiction, Fact, and Threat, Cambridge, MA/London: MIT Press.

7 Stefan Rieger/Armin Schäfer/Anna Tuschling (2021): Virtuelle Lebenswelten. Körper, Räume, Affekte, Berlin/Boston: De Gruyter.

8 Cf. Oliver Grau (2004 [2001]): Virtual Art. From Illusion to Immersion, Cambridge, MA: MIT Press.

to present spectacularly otherworldly realms, which originated from a strong emphasis on the liberation from bodily limitations[9] and on the sovereignty of the mind.[10] Such references increased during the 1990s, culminating in the creation of entire alternative or parallel worlds, including escapist virtual realities, which attracted interest in both art and theory.

Today, digital worlds no longer exist separately but have become embedded in everyday life. Subsequently, the deepened integration of VR into contemporary art today is not primarily of interest to us in terms of its consolidation as a museum art form. Rather, we regard VR experiences that are exhibited in art contexts as another symptom of their arrival in society. In this context, the integration of VR technology is specifically linked to the possibility of questioning the normalization and taken-for-grantedness of the technology.[11] From this point of view, the artistic appropriations of VR serve both functions: they explore the technology's potential as an artistic medium (compared to, and even intertwined with, painting, sculpture, performance, and installation art), while also addressing the hidden agendas of its socio-economic uses and the techno-imaginaries associated with virtuality. Thus, a coherent discussion of VR in art requires an understanding of various frameworks and temporalities. Even though the technology has changed significantly in recent years, some of the promises and fears that emerged in the 1990s still play a role in how VR is perceived today.

---

9 The vision of a disembodied space is expressed primarily in activist and artistic manifestos, including not only John Perry Barlow's *Declaration of the Independence of Cyberspace* (1996) but also, for example, the Ars Electronica exhibition catalogue entitled *Digital Dreams. Virtual Worlds* (cf. Georg Hattinger/Peter Weibel (eds.) (1990): Ars Electronica 1990. Digitale Träume – Virtuelle Welten, Linz: Veritas). However, the works speak their own language, and the theoretical conceptualization of mixed realities was emerging at the same time (see Paul Milgram/Fumio Kishino (1994): Taxonomy of Mixed Reality Displays, in: IEICE Transactions on Information and Systems E77-D/12, p. 1321–1329), even though the model of a linear scale from real to virtual still underlies this framework.

10 See Jens Schröter (2004): Das Netz und die Virtuelle Realität. Zur Selbstprogrammierung der Gesellschaft durch die universelle Maschine, Bielefeld: transcript, p. 216ff. Of course, there were also early experiments which, in contrast, addressed the body directly by providing rudimentary hand avatars (*Home of the Brain*), or corporeal interfaces such as the famous diving suit for navigating through the virtual space by regulating the breath in Char Davies' *Osmose* (1995). Some multimodal works shifted the focus from the visible to the audible as in Akitsugu Maebayashi's *Audible Distance* (1997).

11 Cf. Hans Blumenberg (2010): Theorie der Lebenswelt, Berlin: Suhrkamp.

With our primary focus on the increasing use of VR headsets within contemporary art, we must also take the multi-faceted driving forces behind the renewed interest in VR technology into consideration. After the highly productive 1990s, the second boom around 2015 followed the »resurgence of VR,« which Marcus Carter and Ben Egliston date to around 2010. Similarly to the developments taking place in the 1990s, this moment is also characterized by VR's close enmeshment with the gaming and entertainment industries, that already drove the separation from the initially dominant military context before.[12]

What is new since the beginning of the 2010s is the fusion of VR with the realm of social media, exemplified by Facebook (now Meta)'s acquisition of Oculus in 2014.[13] This development forms the basis for subsequent ideas relating to interoperable metaverses.[14] Users are intended to experience these metaverses as immersively as possible with the help of VR technology. This turns the virtual into their working, living, and leisure spaces, which are ideally interconnected in a seamless way. The progressive takeover of key areas of everyday life is in turn facilitated by gamification, which persists as one of the core principles of the metaverse.[15] In the context of making virtual worlds increasingly inhabitable, another important driver behind the new artistic turn to VR since the mid-2010s comes to the fore: the technical innovations in headsets and enhanced tracking capacities. Features such as the Guardian or the passthrough mode ensure the integration of a virtual (gaming) area as a field of action in the middle of our surroundings, e.g., our living room, not only to avoid involuntary collisions between the virtual and the physical space but also to enable complementarity.

The entanglements between art, society, and technology outlined above, which define the field of artistic engagement with VR, have prompted us to conceive of VR as an interfacing phenomenon. Therefore, we want to focus on how VR brings the three contexts of art, games, and everyday life into relation.

---

12 See J. Schröter: Das Netz und die Virtuelle Realität, p. 99ff. and 210ff.

13 M. Carter/B. Egliston: Fantasies of Virtual Reality, p. 10ff. and p. 51ff. See also Lisa Nakamura (2019): Feeling Good About Feeling Bad: virtuous virtual reality and the automation of racial empathy, in: Journal of Visual Culture 19/1, pp. 47–64.

14 Cf. Boris Magrini (2023): Kunst und das Metaverse heute – ein Wettbewerbsfeld, in: Pamela Scorzin (ed.), Mixed Realities. Neue Wirklichkeiten in der Kunst, Kunstforum International 290, pp. 67–75.

15 Cf. Benjamin Beil et al. (eds.) (2025): Gaming the Metaverse, Bielefeld: transcript.

1. VR exhibited

Given the renewed relevance of VR, both the technology and artistic engagement with it are currently balancing various factors: the emancipatory promises of the 1990s, gaming environments and the latest developments in the normalization of virtuality. However, in order to reflect on our existing lifeworlds, artists reinvest in building entire virtual worlds within the headset, which normally exclude the physical surroundings. What fascinates artists is not so much the increasing optimization that mainly drives game developers and facilitates usability for social media metaverses.[16] Instead, they are intrigued by the inherent possibilities of dissociating and reassembling the individual elements of the interconnections between body, technical apparatus, and world. Rather than focusing on the insularity of these virtual worlds, they increasingly aim at greater permeability between virtual worlds and exhibition spaces.

This is where a crucial consequence of exhibited VR comes into play: Since these VR worlds cannot be experienced from the outside, the empty exhibition space calls for a designed environment. This instantly undermines the idea of an enclosed space of imagination and the freed mind dear to much of the VR art of the 1990s. This situation creates the condition for strengthening the manifold transitions correlating the exhibition space with the VR experience. For example, as an extension of a fictional world or as an engine room revealing parts of the infrastructure which sustains the virtual world, to mention only two of multiple options. It is this situatedness within the exhibition space that is essential for our analysis of VR through art. We would therefore like to further elaborate on this approach to exhibited VR, which not only engages spectatorial, spatial, and institutional considerations but also opens up multiple methodological interrelations.

When viewed from an art-historical perspective, VR invokes bodily practices connected to the performing arts and performance based artistic work, for example. In this context, the technical innovations in headsets and enhanced tracking capacities discussed above come into focus. These technologies synchronize the virtual world not only with user's head movements and gaze, but also with whole-body and hand movements. They are designed

16 Contrary to their promises, these optimizations contribute to the confinement of the inner world and tend to compensate real-life encounters in form of hermetic escapism. Ibid, p. 33ff.

to increase freedom of movement, heighten multi-modal sensation, and strengthen the sense of being able to interact with the synthetic environment and surrounding objects. Artists and curators are well equipped to answer to the call to embed VR-experiences in the exhibition space, given the rich legacy of installation art and site-specific practices of institutional critique. Rarely does the exhibition space remain empty, with only a headset suspended from the ceiling, as was the case in the exhibition *Unreal* 2017 at the NRW Forum Düsseldorf.[17] We therefore propose to reconsider artistic practices, especially those involving installation, site specificity and bodily performance, in the context of VR research.

But art-historical references alone fall short when one enters interactive and exploratory virtual worlds shaped by the affordances of game engines. For museum visitors and researchers alike, familiarity with controllers and some knowledge of the gaming sector are necessary. Dynamic image spaces, level structures, first- and sometimes third-person perspectives, lore, game-like interactions, and exploration modes must be experienced, discussed, and analyzed. These do not always provide a consistent form or linear sequence and vary greatly depending on user behavior.[18] Additionally, it is crucial that these artistic virtual worlds not only relate to art-historical tropes and gaming logics but also increasingly incorporate references to everyday life. This requires reflection on how art relates to socially normalized and therefore often unquestioned technologies, as well as to possible alternative uses of these technologies.

17 See the exhibition *Unreal*, NRW Forum, 25.05.-30.07.2017 which featured a virtual extension of the museum building instead, and the recent project: wwwforum conceived as virtual dependance of the NRW Forum. Cf. Alina Fuchte/Isabelle Hamm (2025): »Virtual Wonderlands, Insights from NRW-Forum Düsseldorf's WWWFORUM Project: An Interview«, in Benjamin Beil et al. (eds.): Gaming the Metaverse, Bielefeld: transcript, pp. 247–264.

18 For the controller as an essential object for VR art see Manuel van der Veen (2025): »Virtuelle Objekte. Eigensinnige Controller in der künstlerischen Praxis von LuYang und Mohsen Hazrati«, in: Patrizia Breil/Alisa Kronberger (eds.), Eigensinnige Objekte. Virtuelle Möglichkeitsräume zwischen Aufforderung und Entzug, Bielefeld: transcript, pp. 25–50. See for some methodological consequences resulting from this Manischa Partowi/Annette Urban/Manuel van der Veen (2025): »Screenshot«, in: Patrizia Breil/Florian Sprenger (eds.), Virtuelle Universität – Geistes- und gesellschaftswissenschaftliche Zugänge, Bielefeld: transcript, pp. 383–392.

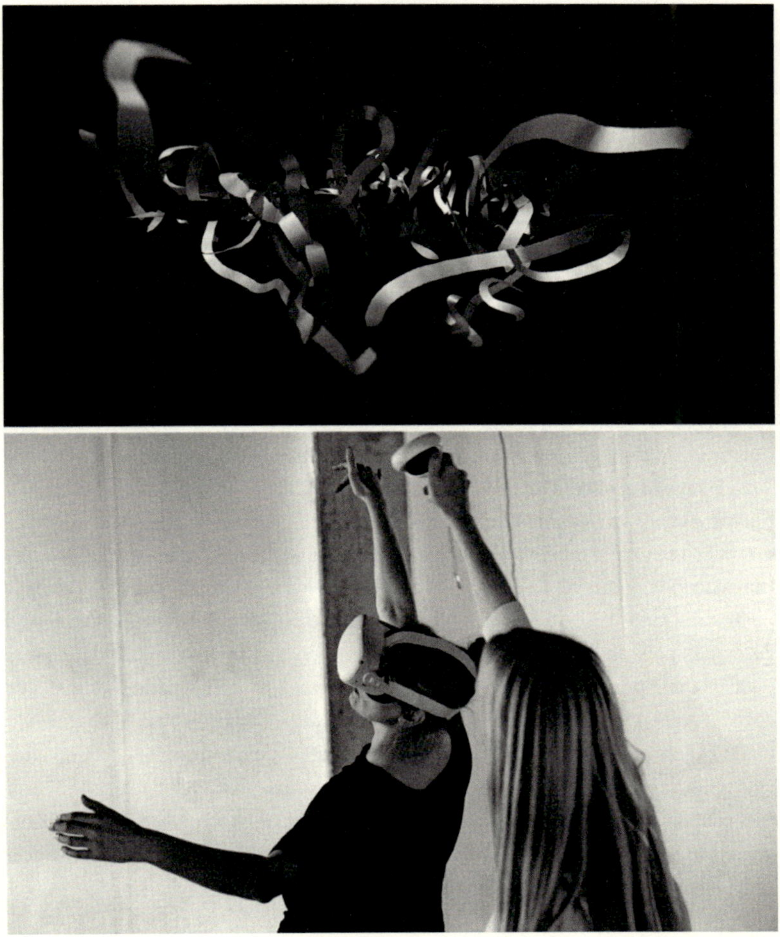

fig. 1: Anna-Carolin Weber in cooperation with Tobias Kopka: I Spy With My Little Eye – Duette für einen Kopf und zwei Hände, 2022 (above: VR-view, below: photography of the performance)

The three areas of art, games, and everyday life overlap in various respects when it comes to VR. What sounds like a complex process can be achieved with simple methods and even without technical intervention: Ann-Carolin Weber engages with this triad in her performance *I SPY WITH MY LITTLE EYE – Duets for One Head and Two Hands* (2022)[19], in which she passes a headset to one person and the controllers to another. While the person wearing the headset perceives the movements activated by the controllers only as loops of color in a dark, otherwise empty virtual space, the person drawing the movement patterns with the controllers is able to observe how the headset wearer reacts with their body to the translations. In this way, the external space is just as relevant as the virtual space; the virtual body is distributed between two agents, and physical movement is translated into a drawing that can be read as a playful dance or social communication. The work could be understood as an experimental intertwining of performance elements with a type of interaction known from games, as well as with possibilities of being in multiple locations at once and drawing on mixed forms of embodiment.[20] This structure also informs our virtual lifeworlds. It is a form of synchronization between various virtual, digital, and physical worlds, achieved through the user's self-reflexive body.[21]

The act of confronting one VR user with a dark, abstract space and equipping the other with controllers allows them to sense each other: once by reproducing the tracked movement and once because without glasses, the surroundings can be perceived directly. This addresses both hermeticism and possible permeability of VR. In Weber's work, the distance between virtual reality and external space is bridged by hacking the device and by instructing the users to shift their perception from purely vision to also imagining and feeling. This may be understood as a distinction between older and newer VR. With the help of DIY techniques, Ann-Carolin Weber indicates a direction that current developments in the high-tech industry are taking. The distance between VR and external space is now minimized through the use of passthrough mode, integrated into the headsets. This feature, known from AR glasses and formerly a safety feature in VR goggles, has only recently become more common in every-

19 *I Spy with My Little Eye* (2022), in: vrdanceclub.de. Online: https://vrdanceclub.de/about-i-spy/ (last access: 17.10.2025).

20 This is connected to a focus on corporeality, which has long been a fundamental aspect of the theorization of the deeply embodied virtual reality, as described by Mark B. N. Hansen in his 2006 findings. See Mark B. N. Hansen (2006): Bodies in Code. Interfaces in Digital Media, London: Routledge, p. 107ff.

21 Julia Reich/Manuel van der Veen (2023): »›a kind of mixed, intermediate experience‹. On the Entanglement of Image and Bodies«, in: Yearbook of Moving Image Studies (Mixed Reality Images), Marburg: Büchner, pp. 92–114.

day life and as an artistic tool; it is therefore of particular importance to us.[22] Passthrough allows the scenery internal to a virtual world to be shown simultaneously with the immediate physical environment. As such, it reclaims a degree of embeddedness of the virtual space within the physical that, until now, only handheld devices have been able to offer.

1.1 Virtual Worldmaking in-between

The technological shift from all-encompassing VR to passthrough VR is a crucial symptom of the transience of technology. In relation to this development, it is the contiguous impact and potential that VR has on art that interests us in particular. Just as virtuality and the lifeworld can no longer be separated from one another, the idea of a hermetic VR is becoming increasingly questionable. In short, it could be said that the wide variety of technical experiments and innovation within VR over the last ten years has led to a complex differentiation of VR's artistic potential that awaits further investigations. Commonly used terms such as »Xtended Reality«, especially the new technical modes referred to as »EMiXAR«,[23] require more precise definitions in order to determine the terms' relation to encapsulating VR. Unsurprisingly, the recent buzzwords referring to mixed realities have affected the previously strictly virtual worlds.[24] As discussed above, these virtual worlds are nowadays casually addressed as interoperative metaverses,[25] where worldbuilding is often considered a trans-

22 Cf. Julia Reich/Annette Urban/Manuel van der Veen (2023): »passthrough. Von Portalen, Durchblicken und Übergängen zwischen den (virtuellen) Welten«, in: Pamela C. Scorzin (ed.), Mixed Realities. Neue Wirklichkeiten in der Kunst, Kunstforum International 290, pp. 86–95.

23 Meaning something like Extended Mixed Xpanded Augmented Reality. See online: https://www.emixar.com/ (last access: 15.10.2025). The X maybe also refers to Crossing, see Paulien Dresscher/Nanna Verhoeff: »XR. Crossing and Interfering Artistic Media Spaces«, in: Larissa Hjorth/Adriana de Souza e Silva/Klare Lanson (eds.), The Routledge Companion to Mobile Media Art, New York: Routledge 2020, pp. 482–92.

24 Cf. P. Scorzin: Mixed Realities.

25 See the exhibition *Collective Worldbuilding. Art in the Metaverse*, 03.06.-13.08.2023, Haus der Elektronischen Künste Basel; Sabine Himmelsbach/Boris Magrini (eds.) (2023): Algorithmic Imaginary. Art on the Blockchain and in the Metaverse, Ex. Cat., Basel: Christoph Merian Verlag; B. Magrini: Kunst und das Metaverse heute; T. Sauerländer: Unleashed Utopias.

medial practice.[26] In this context, the lifeworld significance of virtual worlds is mainly argued on the basis of the participatory involvement of gamer communities and of fans following the connecting points of transgressive storyworlds.[27] Alternatively, it is tied to virtual environments and »free social gaming platforms« that function as »stages for performing online identities.«[28]

In the search for a terminology that captures the currently prevailing mixed conditions more precisely, neologisms such as the »phygital« have emerged.[29] This term was foreshadowed by the »analogital,« which can be traced to Marshall McLuhan,[30] and to a perspective foregrounding the fundamental hybridity of electronic media and transitions from »old« to so-called »new« media in general. Nevertheless, this fundamental mixedness should not lead us to dissolve into a vague de-differentiation. The hermeticism in VR, the phenomenon of an encapsulated world within an image space, still seems to attract artists. At the same time, many VR experiences today – like the one developed by Ann-Carolin Weber – exclude their direct surroundings, but only at first glance. By connecting these divergent tendencies, we could propose that VR's potential is, first, to create some sort of enclosure or shelter, not solely reserved for the dystopian, corporatized spaces of the metaverse, but for the freedom to imagine alternative, even auto(-reproductive) worlds; and, second, to design the transitions between these worlds and the exhibition space.

---

26 Julia Stoschek Foundation (ed.) (2024): Worldbuilding. Gaming and Art in the Digital Age, Ex. Cat., Berlin: Hatje Cantz Verlag.

27 Cf. among others Krzysztof M. Maj (2015): »Transmedial World-Building in Fictional Narratives«, in: IMAGE. Zeitschrift für interdisziplinäre Bildwissenschaft (Special Issue 22) 11/2, pp. 83–96, https://doi.org/10.25969/mediarep/16492. And Patrick Rupert-Kruse (2017): »Transmedia Storytelling oder Erzählen in Medienökosystemen«, in: Jahrbuch immersiver Medien 9, Marburg: Schüren, pp. 6–9, https://doi.org/10.25969/mediarep/18132, who refer back to Henry Jenkins (2006): Convergence Culture. Where Old and New Media Collide, New York & London: New York University Press.

28 Andrea Shinyoung Kim: Virtual Worldmaking. A Phantasmal Media Approach to VRChat, p. 10 and p.12, online: https://cms.mit.edu/wp/wp-content/uploads/2022/03/553835918-Andrea-Kim-Virtual-Worldmaking-A-Phantasmal-Media-Approach-to-VRChat.pdf (last access: 14.10.2025)

29 Pamela Scorzin (2023): »The ›Phygital‹ as the Virtual Real: The Role of Mixed Realities in Contemporary Art«, in: Yearbook of Moving Image Studies (Mixed Reality Images), pp. 115–135.

30 Cf. Verena Kuni (2015): »F (ANALOGITAL)«, in: Daniel Kulle et al. (eds.), Post-digital Culture, online: http://www.post-digital-culture.org/kuni/ (last access: 15.09.25).

While the division of VR devices among users, as in Weber's example, is more of an exceptional case, the convergence of virtual worlds with the physical exhibition space has become a common strategy in the arts. An example is the VR experience *[Posthuman Wombs]* (2022) by Anan Fries. On the one hand, Fries' work requires a form of encapsulation in order to convey the imaginary of a community in which non-binary people can take on and share the task of bearing children. This encapsulation is reinforced by the spatial impression of a world that likewise envelops users like a pastel-colored womb (fig. 2).

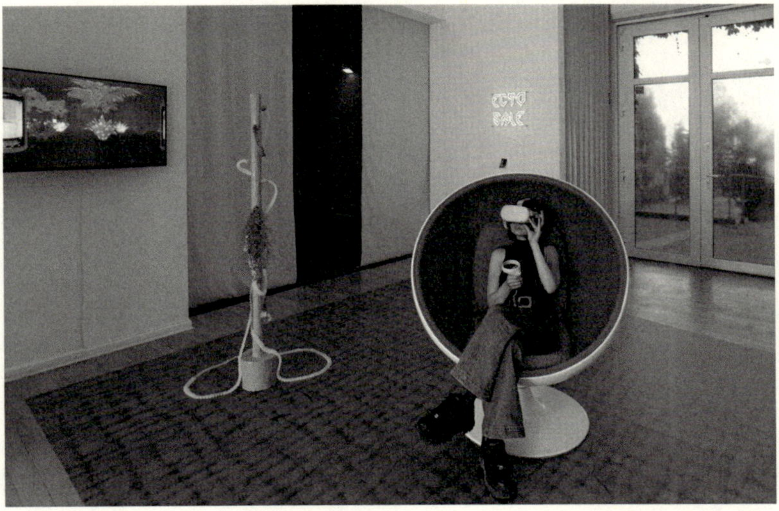

fig. 2: Anan Fries: [Posthuman Wombs], 2022, installation view: Unleashed Utopias. Künstlerische Spekulationen über Gegenwart und Zukunft im Metaverse, group exhibition of VR-Art Prize, Haus am Lützowplatz, 09.09.-04.11.2023

On the other hand, this encapsulation already contains moments of transition. In general, such transitions can be brought about by a wide variety of means and methods, including assembling virtual utopias from scanned fragments of our daily surroundings, thus questioning the relationship between life and alternative worlds. Equally relevant are references to site-specific installation practices that build bridges to the virtual environments or involve experimentations with different modes of experience, thereby creating intersections between exhibition and game spaces. Most of these ways of cross-referencing inner-worldly and physically existing environments, bodies, and objects can be found in Anan Fries' VR experience as well. This begins with life-like body-scans of the artist and their companions, turned into inhabitants of this posthuman future scenario. It continues with the synthetic uterus applied to their body and finally scaled up to a giant IT-bag that cradles the offspring as a literal body accessory. This bag is a transferable shelter that makes care work shareable and a commodity fetish at the same time. Visitors can interact with a digital version of the bag using an AR tablet, and this interaction is integrated into a larger installation setting that includes a ball chair for the VR user.

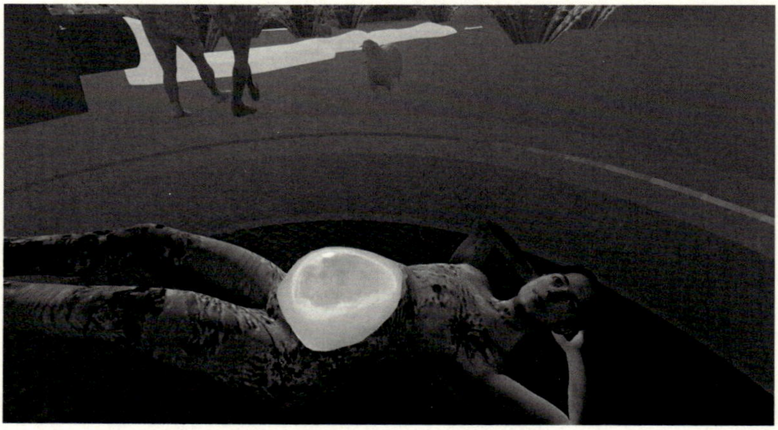

fig. 3: Anan Fries: [Posthuman Wombs], 2022, Screenshot

These multimedial transitions between in-game phenomena, exhibition space, and lifeworlds – both within VR and in physical space – significantly inform the VR experience right from the beginning. As soon as the headset is on, we find ourselves in a cave-like environment where a set of screens plays a tutorial on how to wear the device and operate the controllers. The tutorial gently guides us into the virtual world. The built-in how-to sequence signals that care is being taken for the users, especially for those who are unfamiliar with VR devices.[31] The prelude invites visitors to freely explore the environment (surrounded by smiling pregnant people) and additionally includes explicit hints to different, more interactive sequences, where users can select various feeds scattered across huge mobile-phone screens for closer viewing. Thus, the artwork integrates parts of the social media iconosphere as another crucial element of our everyday life. The virtual experience also offers an inner-worldly diegesis that in many parts consists of a form of guided immersive storytelling reminiscent of VR films (Fries' work is itself described as a VR essay).

To summarize, the example by Anan Fries provides a few central indications of why it is often not sufficient to be concerned with VR as a narrative »hypermedium«[32] and as a remediation, for example, of cinema.[33] It is also necessary to better understand how the invitation to explore such virtual worlds naturally adapts specific affordances and modes of interaction design. First developed and tested in the context of gaming, these technologies now permeate our everyday actions and seamlessly intertwine with the habits of online communication and sociality.[34] Confirming the observation that VR in contemporary art figures between an encapsulation from and a transition to its worldly surroundings, we take this as a starting point for reexamining the relationships between everyday life, gaming culture, and art-historical traditions as well as exhibition practices, as they are mediated specifically through VR. Rather than looking at art as a derivative of VR, we propose to broaden the understanding of VR as an artistic medium.

31 Cf. Christian Bunnenberg et al. (2025): »Von Empathie zum Caring, VR als soziale Konstellation«, in: P. Breil/F. Sprenger (eds.), Virtuelle Universität, pp. 471–483 (forthcoming).

32 Patrick Rupert-Kruse (2024): »Panorama – Sphäre – Skulptur. Versuch einer Narratopologie virtueller Realitäten«, in: Lars C. Grabbe/Id./Norbert M. Schmitz (eds.), Neue Erzählformen in dynamischen Bildtechnologien. Formprobleme zwischen Populärkommunikation und autonomer Kunst, Marburg: Büchner, pp. 83–107.

33 Jay David Bolter/Richard Grusin (2000): Remediaton. Understanding New Media, Cambridge, MA: MIT Press.

34 See Ksenia Fedorova (2020): Tactics of Interfacing. Encoding Affect in Art and Technology, Cambridge, MA: MIT Press.

1.2 Installation Settings of VR art

In addition, the emphasis on exhibited VR as we understand it here shifts the focus away from the issues of easy accessibility, which are primarily enabled by desktop-based WebXR as a location-independent infrastructure. The questions of alternative, non-hierarchical forms of distributing and simultaneously producing digital art that deliberately bypass traditional art institutions have instead been at the center of net art discourses, and they still are with regard to online curating.[35] Web-XR experiences are generally not a frequently used format in art, due to the (still) low distribution of VR headsets in private households. Nevertheless, there are interesting experiments using cardboard devices, WebXR-based navigation via WASD keys, or the open-source kit for virtual rooms made available by Mozilla Hubs until recently.[36]

In contrast to the open gesture of inviting online audiences, artists working with HMD-based VR in museum spaces frequently invest in a single-user mode of reception, figuring the experience as an intimate situation. To this end, they use installation-based means such as Anan Fries' ball chair, a bed[37], a carpet[38], a bumper car[39], self-made machines[40], or a bathtub as placements

35 Annet Dekker (ed.) (2021): Curating Digital Art. From Presenting and Collecting Digital Art to Networked Co-Curation, Amsterdam: Valiz. Or Joasia Krysa (ed.) (2006): Curating Immateriality. The Work of the Curator in the Age of Networked Systems 3, New York: Autonomedia.

36 Cf. for the first option the exhibition *HOLE* by New Scenario that Annet Dekker is analyzing in her contribution to this volume, and for the last option the numerous exhibitions realized by the platform peer-to-space directed by Peggy Schoenegge and Tina Sauerländer: https://www.peertospace.eu/ (last access: 15.09.2025).

37 Cf. Manischa Partowi (formerly Eichwalder) (2024): »Embracing Otherness? A VR Body Hack by Morehshin Allahyari«, in: AN-ICON. Studies in Environmental Images 3/1, pp. 119–132, https://doi.org/10.54103/ai/23206.

38 See, for example, in the VR installation *Palo Alto* (2017) by Banz & Bowinkel. Cf. also Annette Urban (2023): »Escape. Weltbezug und Weltentzug in Kunst mit VR«, in: Lilian Haberer/Karina Nimmerfall (eds.), Mouvement/Movement. Festschrift für Ursula Frohne, Munich: edition Metzel, pp. 99–104.

39 See, for example, the VR installation *Hysteresis* by Dani Ploeger. See also Jens Schröters contribution to this volume.

40 Akwasi Bediako Afrane *TRONS 'R' US*. For the analysis of objects in VR art see Manuel van der Veen (2025): »Headset – Maske«, in: Zeitschrift für archäologische Aufklärung 4, Bielefeld: transcript as well as Id. (2025): »Objekte«, in: Manischa Partowi/Annette Urban/Id. (2025) (eds.): VR in der Gegenwartskunst. Ein Living Catalogue, online: www.vr-in-gegenwartskunst.de (last access: 15.08.2025).

for the VR user in order to underscore the technical shielding of the apparatus. Whereas the strong affective response triggered by this twofold envelopment underpins the much-criticized idea of VR as an »empathy machine,«[41] this aspect of the technology can also be seen to reveal its darker side in various artistic examples. This is particularly true when discomfort and latent distress arise from being transported – or rather, exposed – to a world and its hybrid creatures that loom oppressively and uncomfortably close to the user, as demonstrated by Morehshin Allahyari's VR installation *She Who Sees the Unknown: Kabous, the Right Witness, and the Left Witness* (2019).[42]

Against this background, limiting the scope to exhibited and HMD-based VR art turns out to be a fruitful approach for our art-historical research project. Our aim is to address the contemporary state of virtual art in times of virtual lifeworlds, a good 20 years after the first relevant art historical treatises on this topic.[43] Our approach of situating the VR artwork within the context of an exhibition not only elicits the encapsulating tendencies of VR, but also the multiple modes of transition that artists create to connect the virtual with the physical space, as outlined above. Furthermore, this perspective brings entirely new objects into view: In order to counter the curatorial challenge that VR art poses to individual, enclosed users and to support the gathering and co-presence of museum visitors, artists have introduced installation elements like a second screen.[44] In our analysis, the second screen has, in particular, turned out to be a central element that artists draw on not simply to provide the museum's audience with insights into what is happening inside the solitary VR experience, but to extend and complicate the work's spatial and conceptual logic. Many artists transform this necessary element, which first and foremost improves VR's suitability for exhibition in the museum space, into a multi-layered installation device. For example, the external second screen can be repeated inside the VR experience, functioning like a reverse window that inserts a live view of museum visitors into a virtual, extraterrestrial world, such as in *Mercury* (2016), a VR installation by the artist duo Banz & Bowinkel.[45] This second screen can

41 Cf. the contribution by F. Cavaletti/P. Conte/A. Pinotti to the volume where the extensive discourse on this topic is concisely summarized.

42 Cf. M. Eichwalder: Embracing Otherness? See also Manischa Partowi's discussion of Jordan Wolfson's *Little Room* (2025) within this volume.

43 Cf. O. Grau: Virtual Art.

44 See fig. 1 in the contribution by Manuel van der Veen to this volume. See also Id. (2025): »Second Screen«, in: M. Partowi/A. Urban/Id. (eds.), VR in der Gegenwartskunst.

45 Cf. also the contribution by Ursula Ströbele to this volume.

function as a mirror image that transforms the avatarless embodiment of the VR user into an empty shell.[46] Or it can, as in Patricia Detmering's VR installation *Aporia* (2021), separate the VR-internal world experience from the wall-filling projection of the second screen by introducing two different perspectives onto the same virtual landscape.[47] This strategy stimulates an exchange between the no longer completely isolated immersive participant and the bystanders as they compare their respective worldviews.

In order to systematize the transitions outlined so far, which demonstrate how exhibited VR is becoming permeable from within, this research project has benefited from looking at its references to the world, the body, and objects.[48] Under this condition – namely, a virtual reality that opens up from within and is situated in exhibition spaces – we would now like to take a closer look at the aforementioned interfacings between the fields of art, computer games, and everyday life. These interfacings do not coincide with the transitions between physical and virtual space, but rather require different modes of experience and approaches to interpretation, as well as references to a broader field of the virtual that goes beyond contemporary art.

2. Interfacing

Our previous discussion has shown that it is necessary to consider the shift towards now widely used technologies of virtuality in order to better understand the changing conditions that have led artists to return to VR. We propose a framework that moves beyond the interfaces that are decisive for access to virtual experiences and often specifically designed by artists as hybrid objects, in favor of rethinking VR as a phenomenon of interfacing in itself.

---

46 Cf. the contribution by Manischa Partowi to this volume that analyzes VR installations by Theo Triantafyllidis and Jordan Wolfson.

47 Cf. Annette Urban (2026): »Insuläre Topografien und welthafte Ganzheit. Künstlerische Auseinandersetzung mit Landschaft in VR«, in: Anna Polze/Manuel van der Veen (eds.), Virtuelle Landschaften. Raumerkundungen an der Grenze des Screens, Bielefeld: transcript (forthcoming).

48 Within CRC-project C03 *Virtual Art*, the members of the research team are each working on their own subprojects: Manischa Partowi is focusing on aesthetic experience in VR art, Manuel van der Veen is looking at specific object constellations, and Annette Urban is examining the question of their references to the (life)world.

We thus convert a technical feature of VR – one that is based on transitions in terms of exhibition, production, and conceptual definition – into a methodological approach. We thereby propose transitions between perspectives from different disciplines as particularly effective for the analysis of VR art. Given this analogy, VR as a technique of mediation between very different areas does not necessarily have to be broken down into a multitude of functionally differentiated fields. Rather, the technology can be meaningfully systematized for us through the triad of everyday life, gaming culture, and art. As an anthology title, we have chosen *Virtual Reality Exhibited* not only to indicate our subject matter of VR in the exhibition space, but also because we wanted to signal our wish to ex-pose VR art to different perspectives and frames of analysis. Therefore, we have chosen to focus on the processes and relationality indicated by the verb form *interfacing*[49], and explore how VR also mediates between different arts, techniques, and forms of cooperation that are just as fundamental as the connection of different worlds, bodies, and objects. We understand the concept of interfacing as a framework to address environmental and situational embeddedness and world-building, embodiment, object-relatedness, and the orientation towards recipients. This perspective has proven to be productive as a point of focus for initiating an exchange between the various disciplinary experts on VR. Before we proceed to go into more detail about the workshop's findings, our framework built around interfacing can be illustrated with one final example from our body of work.

DiMoDA 4.0 Dis/Location (2022) is a paradigmatic example of VR as an interfacing technology. The project is a virtual exhibition that travels between physical gallery spaces as well as the individual's desktop. In the exhibition space, it can be experienced via an HMD, and on a personal computer, it can be downloaded from the developers' website and experienced in 2D. *The Digital Museum of Digital Art* is conceived as a »pioneering virtual institution, dedicated to commissioning, preserving, and exhibiting cutting-edge VR artworks.«[50] It was founded by the artists and developers Alfredo Salazar-Caro and William Robertson in 2013. Up until now, there have been four editions of *DiMoDA*, each consisting of a unique virtual environment with a museum-like architecture that hosts an exhibition of VR artworks. In *DiMoDA 4.0 Dis/*

49 See K. Fedorova (2020): Tactics of Interfacing, and Iris van der Tuin/Nanna Verhoeff (2022): Critical Concepts for Creative Humanities, Lanham: Rowman & Littlefield.

50 See the handout for the exhibition of *DiMoDA 4.0 Dis/Location* at Bannister Gallery, Rhode Island, 6.-28.10.2022.

Location, the developers collaborated with the architect Ayman Tawfeeq and the acclaimed curator for digital art Christiane Paul. Both the curatorial concept and the exhibition design are interwoven with the theme of *Dis/Location*. The artworks, such as Ricardo Miranda Zúñiga's *Desplazados* (2021), Tamiko Thiel's *Atmos Sphaerae* (2021) and Banz & Bowinkel's *Grid* (2021), negotiate the experience of dislocation from various perspectives, ranging from gentrification and migration contexts to human-machine miscommunication issues to an abstract and global perspective on human imaginaries and ecological atmospheres. Constituting separate virtual experiences within the main exhibition environment, each of these artworks proposes its own approach to VR as an artistic medium. As a close study of the artworks would exceed the scope of this introduction, we will instead look at the dis/locating dimensions of the virtual exhibition infrastructure to exemplify our understanding of VR as an interfacing technique.

Let us start with the exhibition in the sense of a physical gallery space. Here, the spatial installation of the VR technology creates an intersection between the physical and the virtual exhibition space, which allows for a transition of objects and experiences between these spheres.[51] Positioned in the middle of the physical exhibition space, the VR viewer is surrounded by an audience, who can follow the viewer's virtual experience and POV in real time via a monumental video projection on one of the exhibition walls. As such, the audience and the viewer share both exhibition spaces with each other. At the same time, each viewing experience is configured by the particularities of each of the spaces: Within the physical exhibition space, the audience follows the embodied VR experience as if watching a film. Their mode of experiencing is more decentralized than in VR, which is enhanced by the prominent display of the VR hardware. Like a sculpture, the gaming PC is mounted onto the wall, demonstrating computational power as a crucial structural difference between virtual and physical exhibition space. This display of the hardware highlights the general infrastructural condition of VR as a digitally simulated image space. Next to it, there is an object that migrated from within the virtual into the physical space. It is a 3D-printed object featuring shapes of human heads, abstracted into a polygon structure. Materialized, mounted onto a pillar, and illuminated with a spotlight, this object functions like a sculpture. It can be observed from various vantage points within the physical space, and it links back to the virtual

51 Cf. A. Urban: Über Portale, Sphären und andere eigensinnige Objekte, p. 131ff.

space. In VR, the structure's function is not that of an object, but that of a large building to be partially perceived from outside and explored from the inside.

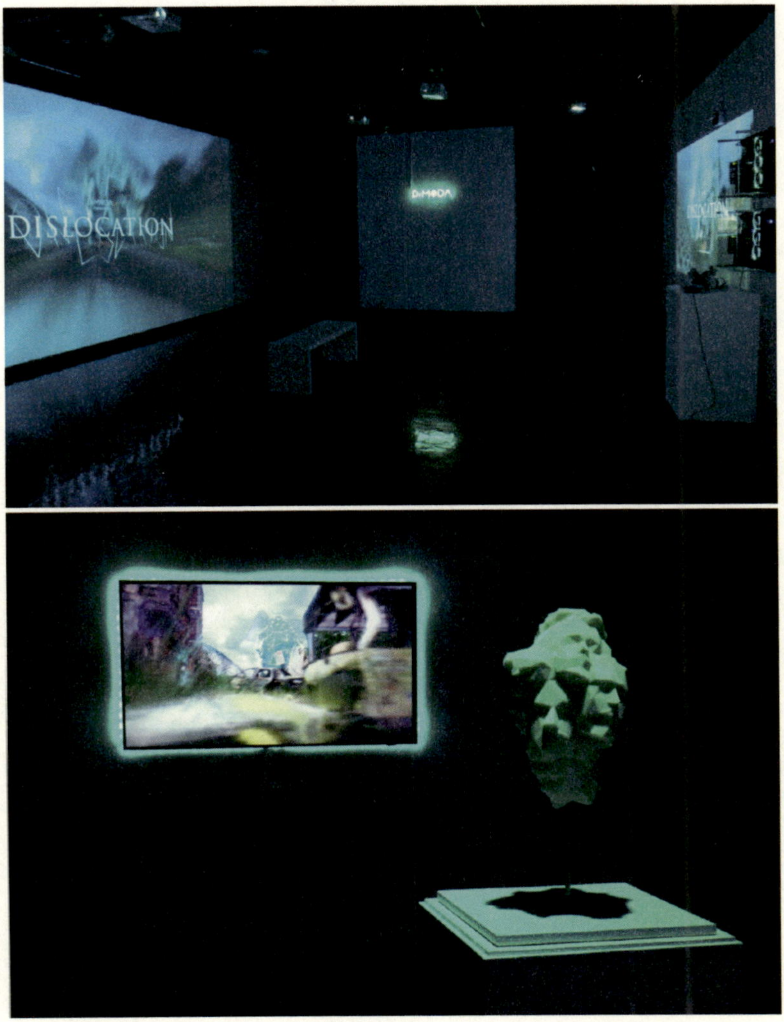

fig. 4: DiMoDA 4.0: Dis/Location, 2022, installation view: Bannister Gallery, 2022

This flexible scalability, which ensures that the built lifeworld, an artifact, and human effigies collapse into one another, is worth further discussion within theories of architectural design, where VR is considered a particularly suitable medium, as it allows users to enter a rendered visualization of a designed space and engage with this image intentionally.[52] In the virtual space of *DiMoDA 4.0*, the building-like structure constitutes the architectural shell that houses various single VR artworks. Extracted from its virtual embeddedness, this shell structure can be understood as a document of the production process itself. As an interface between the virtual and the physical exhibition space, this installation setup offers a reflective approach to the virtual situation and allows for negotiating the different conditions of spatial, material, and experiential stability in VR and IRL.

In the VR mode of *DiMoDA 4.0*, however, the relation between physical and virtual space is disrupted. The viewer is displaced from the gallery space and relocated to an urban environment. With empty and partly demolished buildings, wrecked cars overgrown with vegetation, and an eerie, ghostly atmosphere, the scenery is reminiscent of a post-apocalyptic scenario from a videogame (such as *Fallout 4* or *The Last Of Us*). Further research could help clarify what forms of interfacing are at work in the realistic 3D replicas of buildings from New York neighborhoods that serve as the game's backdrop: How do they relate to photogrammetry-based methods of game production? Moreover, how does the work relate to sociological approaches to neighborhood research, considering that Zúñiga's artwork includes a field research perspective in the form of interviews with residents? The appearance of the houses and the distinctive yellow cabs also evoke the urban space of the megacity, thereby positioning the virtual exhibition scenario somewhere between gameworld and lifeworld.

Unlike in real life, the viewer is not using their body to explore a space but is instead required to know how to operate the controllers to move through it. When navigating the virtual space, the viewer must aim in one direction, pull the trigger on the controller, and release it to land at the desired location. This form of teleport-based, discontinuous movement resembles the aiming mechanism used in many video games, which reinforces the association with a

52 Cf. Philipp Reinfeld (2022): »Performatives Entwerfen in Virtual Reality: Körperaktive Hybridisierungen zwischen virtuellem und realem Raum«, in: Kassandra Nakas/Id. (eds.), Bildhafte Räume, begehbare Bilder, Leiden: Brill/Fink, pp. 27–51, here: p. 30, https://doi.org/10.30965/9783846767238_004.

gaming environment. Yet because there is no hidden quest or task prompting the viewer to take action in order to encounter the artworks, attention eventually shifts away from exploring the deserted surroundings and toward the specific location highlighted within the scene. There, a polygon-shaped object towers between empty houses and is marked with bright blue neon tubes. Like the direction signs that lead visitors from a gallery's lobby to the exhibition space, these signs are intended to be understood literally by the viewer in order to locate the actual artworks. In terms of the orientation of the viewer, the logic of the physical gallery infrastructure seeps into the virtual space and intersects with the mediated view of gameworlds, which is characterized by a »distinctive sensitivity for signs [and] for functional relations.«[53] Based on these observations, we find it promising to pose research questions from different perspectives: How do interaction principles from gaming intermix with spatial relations opened up through curation, especially since computer games always have something to do with space?[54] Furthermore, to what extent does the design of the exhibition world refer to a lore that evokes the social and political dynamics of dis/location?

In previous editions of *DiMoDA 1–3*, the virtual gallery space resembled traditional forms of museum architecture, ranging from classical prototypes to modern futurism. These architectural citations underscore the developers' aspiration to establish *DiMoDA* as the »finest holding for twenty-first century Digital Art in the World,«[55] able to measure up to the gold standards of traditional architecture for art. In *DiMoDA 4.0*, however, these references are absent. Instead, the theme of dis/location is translated into the design of the vir-

53 Marc Bonner (2018): »Gekerbte Wildnis – Inszenierungen vermeintlich unberührter Umwelt in digitalen Spielwelten«, in: Paidia. Zeitschrift für Computerspielforschung (28.02.2018), p. 8, online: https://paidia.de/die-gekerbte-wildnis-inszenierungen-vermeintlich-unberuehrter-umwelt-in-digitalen-spielwelten/ (last access: 15.09.2025).

54 »Even before game studies was a recognized research field, space has been a recurring, core topic of debate on the new cultural form of videogames. [D]ue to the dynamic nature of simulations, space was acknowledged as a constitutive factor of designing and playing games; their sine-qua-non.« See Espen Aarseth/Stephan Günzel/Erik Hoops (2019): »Introduction. Space – The Theoretical Frontier«, in: Id. (eds.), Ludotopia, Bielefeld: transcript, pp. 7–10, here: p. 7.

55 According to the statement of the developers on *DiMoDA 2.0: Morphé Presence*, 08.08.2016, online: https://transfergallery.com/wp-content/uploads/2016/11/DiMoDA-2.0-Press-Release-Updated.pdf (last access: 15.09.2025).

tual environment, which draws on the familiarity of a real-life setting while at the same time remaining significantly detached from ordinary experience. The virtual exhibition still takes place within a building – and not, for example, in the »public« space of the decaying cityscape. Yet the gallery's architecture is more closely integrated into a cyberspace vision of the virtual. Its metallic wire-mesh structure resembles scrap metal in the urban landscape and frames the neon-lit building as a futuristic recycling project, adding a steampunk atmosphere to the virtual environment.

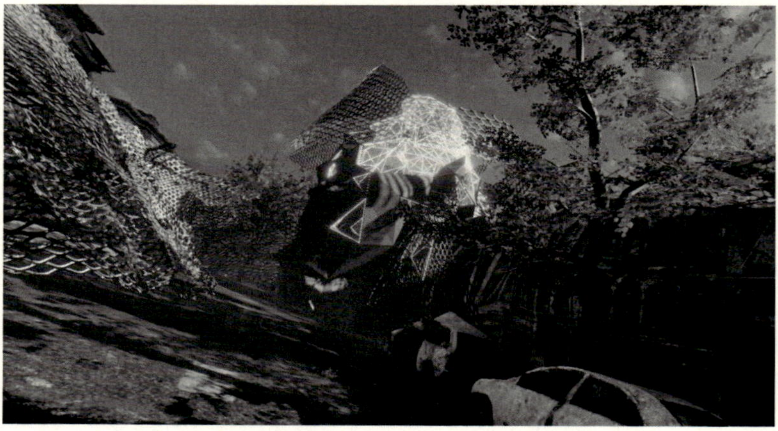

fig. 5: DiMoDA 4.0: Dis/Location, in game screenshot

Upon entering the alien building, the viewer can ascend and descend on a helix-like ramp to reach the artworks. Each artwork consists of a virtual world in its own right, represented as a glowing sphere hovering at different levels of the building. No wall labels or information about the artworks or artists are attached to these spheres, which frames the encounter as a surprise in the style of typical game mechanics. To actually experience the artwork, gaming experience is an advantage as well: just as the neon light of the building signals that an action should be taken, the viewer needs to position themselves within the liquid-like sphere to activate the »action point« and start the artwork. Entering the sphere means entering the artwork – a logic that resembles moving into the next level in games. While entering requires bodily engagement with the virtual representation of the artwork, leaving requires the use of the controller. By opening a menu, a »non-diegetic augmented reality marker«[56] overlays the virtual scene, presenting several items that explicitly provide navigational options. The viewer can leave the artwork by clicking »back to lobby,« which teleports them into the urban space of the virtual setting, or they can open a text containing background information on the individual artwork as well as the exhibition concept. Since conventions of orientation in traditional museum contexts overlap with those of video games, it is productive to conduct a comparative analysis of navigation in computer games and, for example, exhibition texts. This dis/location – or interface – between art and game-world becomes particularly visible within the context of *DiMoDA's* installation in a physical gallery space.

As a download, the virtual exhibition enters the viewer's lifeworld. This personal mode of display dislocates the viewer from their habitual use of the computer and redirects them into the exploration of the peculiar exhibition space. Navigation within the 2D monitor version follows the digital conventions of moving through space with WASD, as in a desktop-based videogame. Far from »browsing the catalogue with a new interface,«[57] *DiMoDA* functions as a unique virtual exhibition format that merges art-institutional conventions with videogame mechanics. This virtual exhibition tends to emphasize an encapsulation from the lifeworld and the artworld by proposing a game-like counterworld. Nevertheless, its particular physical and conceptual embeddedness within these different reception contexts illuminates its sensitivity to the hybrid constitution of VR and the potential of interfacing various research perspectives.

To summarize the multiple interfacings between art, games, and everyday life occurring here: The virtual exhibition setting of *DiMoDA* addresses a crucial parameter of traditional reception contexts. By creating an immersive virtual setting as an encompassing environment for the artworks, it challenges the basic assumption that the physical exhibition space functions to create distance between viewer and artwork. Within the framework of gaming, this setting can be characterized as a form of worldbuilding, which projects a fictional scenario grounded in a dystopian perspective on our hybrid lifeworld. This game-

56 M. Bonner: Gekerbte Wildnis, p. 10.

57 Annet Dekker (2021): »Curating Digital Art. From Presenting and Collecting Digital Art to Networked Co-Curation«, in: Id. (ed.), Curating Digital Art, pp. 14–33, here: p. 19.

like setting creates a tension between the expectations associated with experiencing art – which traditionally emphasize reflection – and those associated with gaming, which emphasize immersion. Yet, as demonstrated in our discussion of the virtual exhibition environment, the need to adapt to this form of orientation introduces a rupture in the supposed seamlessness of the virtual experience and provides an entry point for reflecting on its constitutive parameters. At the same time, *DiMoDA* embraces a contemporary drive toward immediacy[58] by translating a general expectation of digital interactivity into physical space. However, *DiMoDA* also explicitly references art-specific conventions, such as the placement of artworks within a building structure, signifying a shift in reception and thus a distinction between the everyday and the artistic. It not only creates a physical exhibition structure within VR but also transports this virtual version back into the physical gallery, thus mirroring the relationships between architectures inside and outside VR. As a digital platform for collecting, archiving and exhibiting VR art, *DiMoDA* corresponds to the traditional purposes of a museum, yet proposes a VR-influenced approach to collection display – one that enables the traversal of multiple virtual worlds.

3. The Sections

As this example has shown, focusing on the relations between art, games, and everyday life is a valuable approach for broadening the perspectives from which VR art can be analyzed. Interfacings are integral to various dimensions of VR art: to the ways image, space, and time are constructed; to its modes of reception; to its means of production; and to the broader institutional and sociopolitical contexts of the medium. The framework of interfacing not only allows us to better situate the complex nature of VR art within its genealogy and current developments but also underscores the necessity of integrating research from various disciplines. In this volume, we propose the interfacing of art, games, and everyday life as a shared framework for addressing VR in art from the perspectives of philosophy, game and media studies, art history, and exhibition-making. In doing so, we aim to multiply the methodological approaches and tools for engaging with aesthetic, technological, and practical questions surrounding VR art.

58 See Anna Kornbluh (2024): Immediacy, Or the Style of too late Capitalism, London: Verso books.

The very designation of VR as »virtual reality,« that is a spatial and corporeal medium that stands in a peculiar relation to what we understand and perceive as real challenges a basic understanding of how to situate ourselves within the context of VR art. In its continuous approximation of the real, the medium itself is in constant flux, which makes it necessary to adopt philosophical perspectives in order to grasp its central concepts, experience-related phenomena, and the ways these are addressed in art. Curatorial perspectives provide insight into institutional approaches to VR art and draw attention to the pragmatic and conceptual challenges involved in the technical installation of VR, as well as to the visitor's experience – both of which differ significantly from conventional exhibition traditions. In turn, these dimensions of VR art give rise to new curatorial practices and new spatial and spectatorial concepts, responding to the altered conventions of our contemporary hybrid lifeworld.

Practical experience with VR from production contexts is equally relevant, as it reveals the scope of possibilities, default logics and limitations offered by VR, thus pointing toward future tendencies of the medium and its uses. Media studies can furthermore highlight the complex genealogies, historical virtualities, and contemporary platform-capitalist dependencies associated with VR, which are themselves subjects of artistic inquiry. Game studies and film- and theatre-studies, in turn, constitute a rich framework for aesthetic analysis, as numerous artistic practices across time-based, spatial, digital and non-digital media are often brought together in VR art. With this volume, we propose a methodological integration of these perspectives into art-historical research on VR art. Beyond proposing the tentative cluster of interfacing relations between art, games, and everyday life, we also observe a more overarching thread running through the various contributions: a tension between seriousness and playfulness. This polarity cuts across both the contexts of art museums and entertainment culture; it interfaces historical genealogies of the medium and its established uses within gaming culture; and it constitutes a particular artistic mode of navigating between the virtual and the real, one that is central to art-historical traditions concerned with conceptualizing the aesthetic itself.

Section 1 Art/Games

Genealogies of VR: Sculptures, Environments, Simulators, Magic Circles

The idea of interfacing also broadens the genealogies in which VR can be situated within art history and media culture, extending far beyond references to panoramic images and illusionistic painting. The contributions collected in this section refer, among other things, to the spatiality of objects. Because VR environments are typically produced using 3D-modeling software, objects within them can often be circulated, touched to a limited extent, and experienced in relation to the body. These characteristics evoke a long tradition of sculpture, whose photographic reproduction through photogrammetry has, in turn, developed techniques relevant to the creation of 3D models today.[59] In addition, the ability to design entire rooms relates to traditions of environmental art, which share the characteristic of enclosing the viewer – placing them in a room within a room – and emphasize sensorial, multimodal experience. Finally, among the technical precursors of VR are simulation and simulators, which had already focused on the transition between the virtual and the physical by duplicating aspects of the world.

In examining these genealogies, the contributors to this section explore how VR art often evokes references to high art while also drawing on entertainment and pleasure. In this way, VR adds another mode of engagement to the spatial and artistic constellation. Environmental art opens up a playful zone for viewers, ranging from climbing courses and spatial deformation to dancing. Yet, as the examples in this section demonstrate, the playful aspect of VR art is often contrasted with its use in creating ruptures in the seamless experience, thus highlighting, for example, VR's paradoxical intertwining of entertainment and military technology,[60] or the ambivalent potential of the virtual as a source for (de)constructing our relationship to history.

Sculptural lineages in VR art are examined in **Ursula Ströbele**'s contribution *The Afterlife of Antiquity in Virtual Reality*. She focuses on the role of art-

---

59 Cf. Jens Schröter (2009): 3D. Zur Geschichte, Theorie und Medienästhetik des technisch-transplanen Bildes, Munich: Wilhelm Fink.

60 Ursula Frohne (2016): »Expansion of the Immersion Zone. Military Simulacra between Strategic Training and Trauma«, in: Fabienne Liptay/Burcu Dogramaci (eds.), Immersion in the Visual Arts and Media, Amsterdam/New York: Rodopi/Brill, pp. 214–248, https://doi.org/10.1163/9789004308237.

historical references, such as when 3D scans of ancient sculptures end up in VR worlds. Contrary to a simple idea of translation, the author demonstrates that the sculptures are equipped with additional digital features as they pass through artistic asset libraries and game-engine software. Viewers of virtual worlds are thus challenged to navigate between conventional modes of reception for sculpture and experimental codes of behavior familiar from computer games. Using the counterintuitive example of materiality that, for example, on a virtual Mercury shifts from hard marble to soft rubber during a VR experience, Ströbele points out that the concept of history itself is subject to a certain elasticity through the playful and artistic use of sculptures in VR.

Curating Immersion. The Origins, 1949–1969: *Behind the Scenes and Some Afterthoughts* is the title of **Choghakate Kazarian**'s contribution, in which she reveals the curatorial, conceptual, as well as pragmatic decisions behind a major exhibition on non-digital immersive art, such as environments. She characterizes the 2023 exhibition as a response to popular VR formats like *Virtual van Gogh*, which provide spectacular immersive museum experiences of artworks that originally were not intended to be experienced this way. By proposing a pre-digital notion of immersion »as a totalizing, self-contained, and physically enveloping phenomenon«, Kazarian reflects on the selection of the exhibited artworks as »proto-VR interfaces« and discusses their relational embedding within the institutionalized museum space.

With *On Dani Ploeger's Hysteresis*, **Jens Schröter** demonstrates how different media genealogies intersect in VR art. To experience Ploeger's work, viewers sit in a self-modeled bumper car and then take a high-speed virtual ride in VR, which ultimately ends in a crash. According to Schröter, the bumper car evokes the military use of simulators for accident prevention, as well as the entertainment industry, whose appeal often lies in playfully provoking accidents. In addition to the bumper car, a fog machine forms part of the installation and provides the immersive integration of the VR work into the exhibition space simulating extreme battlefield conditions on the one hand, and evoking the aesthetics of nightclubs on the other, the installation again reveals the tension between seriousness and playfulness in a VR artwork.

Manuel van der Veen explores *The Area of the Virtual* in his contribution and, to this end, presents his concept of *areal space*. This concept is derived from a technical specificity of all XR procedures, namely that they perceive and project space. Beyond categories such as ›virtual‹ or ›real‹, which are often used to define VR as the creation of a virtual world or AR as the superimposition of reality with virtuality, the aforementioned technical specificity allows for a more historical and theoretical connectivity. Therefore, the author examines various works of contemporary art that employ VR and reveal connections to techniques such as the floor plan and the magic circle, which also open up *areal spaces*.

Section 2 Games/Everyday Life

Concepts of VR: Space, Worldbuilding, Presence

Even though the virtual, as in the casual use of digital devices, has established itself as an integral part of our everyday life, the particular qualities and conditions of VR have prevented the technology from becoming ubiquitous. As Marcus Carter and Ben Egliston discuss, the idea of »VR occupying a more mundane role: situated in everyday domestic spaces and used for purposes beyond gaming« is still a utopia, prominently promoted by tech companies.[61] When looking at the inaccessibility of VR in terms of bodily and cognitive disabilities, for example, this vision of VR turns out to be a »libertarian, identity-free, and disembodied fantasy,«[62] rather suited fictitious worlds of computer games than to the ambiguous, messy, and material entanglements of the everyday lifeworld. The shared inheritances with the military and videogames become apparent in the constitution of the viewing apparatus of the HMD as well as in the mechanics of the game engine. On the one hand, this leads to attempts at simulating real-life scenarios. Here, virtual embodiment means an approximation to habitualized forms of bodily orientation required for a technological translation of common conceptions of space and presence into the virtual. On the other hand, VR as a tool for creating worlds encompasses practices of worldbuilding that emerged in the field of game development and proposes networked, web-based approaches to body, space, and time. By focusing on recent developments in passthrough technology, the default setting of the Guardian, as well as on virtual exhibition formats, the contributions in this section contribute to the reconceptualization of the notion of space, the phenomenon of presence, and the practice of worldbuilding.

In the chapter *The Spatial Apriori of Virtual Reality*, **Stephan Günzel** inquires into the spatial conditions of VR, which he unfolds by focusing on the tech-

61 M. Carter/B. Egliston: Fantasies of Virtual Reality, p. 57.

62 Ibid., p. 22.

nical phenomenon of the Guardian system. Originally installed as a security feature, the Guardian allows for simultaneous viewing of the immediate surroundings via video passthrough and an abstract geometric space that serves as the tectonics of virtual »architecture.« When analyzing this partly live-view based partly computer graphics-based combination – or, as Günzel puts it, following Foucault, this »empirico-transcendental doublet« – leads to three insights regarding the spatiality of VR: 1. In contrast to the much-discussed hermeticism, VR is opened up to an empirical perception of space. 2. VR is also subject to a certain spatial conception, which in turn is historically conditioned. The author demonstrates that space is not universal, but rather related to everyday life and regional circumstances, i.e. a specific lifeworld. 3. The author asks how this spatial a priori is reflected in VR. Using various VR artworks and VR computer games, Günzel discusses attempts to undermine this specific VR spatiality and declares that these ultimately do not (yet) escape their own conditionality, which has historically shaped VR.

In her contribution *World[build]ing Online Exhibitions, or Trying to Resist Zombie Curating*, **Annet Dekker** develops an analytic framework for examining virtual spaces in online exhibitions. By focusing on the curatorial practice of the artist duo New Scenario, she examines the distinction between the more static concept of digital worldbuilding, known from traditional computer-game design, and the relational, fluid, and speculative approach of worlding. Emphasizing a conceptual shift from a virtual »world-as-object« towards a »world-as-practice,« Dekker proposes a reconceptualization of VR as a response to web-based phenomena such as the inherently co-creative character of digital space, the coexistence of multiple modes of experience, as well as a distributed form of agency. In her writing, Dekker incorporates the simultaneity of multiple viewpoints that continuously complement each other and suggests a unique methodological approach to situating the position of the researcher within the worlding of New Scenario's exhibition projects.

Chris Salter questions historical concepts of presence that are prevalent in the imaginary of VR technology. In his article *Rethinking Presence in the Age of Mixed Reality*, he develops a hardware-based approach to the conception of presence, which becomes radically challenged by new forms of passthrough technology. By focusing on the genealogy of the »viewing instrumentarium« of the HMD, Salter reviews central VR literature and contrasts their notion of presence as a hermetic illusion of space with a different conception of presence from theater studies. Instead of conceiving presence and its corresponding phenomenon of immersion on the basis of a »suspension of disbelief,« he argues for an understanding of presence as a rupture, a »shock« that highlights the experience of immersion as artifice. Salter discusses the effects of passthrough based technology as a process of »defamiliariz[ation]« of our exposure to the technology and thus proposes a reconceptualization of presence as an experience of the mediated nature of the virtual space.

Section 3 Everyday Life/Art

Modes of VR-Reception: Corporeality, Empathy, Immediacy

As becomes visible in the contributions of the first two sections, the extensive debates on immersion only partially cover the attitudes and affordances required for reception in VR. The fundamental ambiguity of the lived-body experience must be acknowledged, which also means that a »responsible subject«[63] is implied in VR. Conversely, media-infused immersion has become a general condition of the lifeworld, one that is also attributed to architecture and the ubiquitous need to perform.[64] This results in an overall aestheticization of the everyday, which oscillates between disconnecting from the physical world and a regained playfulness.

The contributions collected in the last section of the volume participate in this change of perspective by inviting a closer look at the active adjustments and conflicting expectations that shape the immediacy of embodied VR experience. These conflicts and the resulting negotiations can take place on a very basic existential level, where the body acts as an interface for self-regulation in the face of the uncertain correlations with the world. In an intensified form, this is what can be played out in VR. In other cases, the clash of different institutional settings, media dispositifs, and audience expectations rooted in the art field, can be used for basic exercises in virtuality. They don't primarily aim at improving users' interactive abilities. Instead, they deliberately train

---

63 Zeynep Akbal (2023): Lived-Body Experiences in Virtual Reality: A Phenomenology of the Virtual Body, Bielefeld: transcript Verlag, p. 73ff., https://doi.org/10.1515/9783839466766.

64 Cf. Laura Bieger (2015): Ästhetik der Immersion. Raum-Erleben zwischen Welt und Bild. Las Vegas, Washington und die White City, Bielefeld: transcript, and Doris Kolesch (2021): Ästhetik der Immersion, in: Georg W. Bertram/Stefan Deines/Daniel Martin Feige (eds.), Die Kunst und die Künste. Ein Kompendium zur Kunsttheorie der Gegenwart, Berlin: Suhrkamp, pp. 422–441.

them to collide with other immaterial entities in virtual space or to endure being processed by an intimidating technical apparatus. These examples can be categorized as a subtle interfacing of everyday virtual (trans)actions that are otherwise highly functional and effective in gaming and simulation but here are reclaimed and re-distributed by art.

Ksenia Fedorova and **Jose Hopkins B**. have combined their corresponding thoughts under the heading *Simulation as Attunement: Metastability, Coherence, and Embodied Orientation in Virtual Reality*. Their argument against understanding VR as self-contained, which they correlate with the supposed seamlessness of simulation, is based on the fundamental anchoring of the virtual in the body – or rather, in the constant bodily reconfiguration that characterizes the processual condition of the self. By incorporating their own findings on proprioception, interfacing, and the concept of metastability, they offer a new approach to the line of thought about virtuality rooted in sensory physicality and actual perception. Fedorova and Hopkins develop an understanding of metastable interfacing across different realities which they apply to various exercises with VR in relation to dance and gamified testing grounds for colliding bodies, as well as to avatar encounters within social VR. This approach promises insights into how meaning-making functions through the temporary formation of coherence. By clarifying what it means to feel »real« in the virtual world and how to develop a truer virtual self, the authors propose that this coherence is what makes the XR-self livable.

In their contribution *Playing Against VR. Immersive Environments between Hypermediation and Remediation*, the co-authors **Federica Cavaletti**, **Pietro Conte,** and **Andrea Pinotti** also address the truthfulness of the virtual, but from an opposite perspective. They question the claim to truth associated with the transparency of the medium and its illusion of place, which they situate in the fields of immersive journalism and VR documentaries – both important motors of VR's everydayness. In a second step, these claims are confronted with artistic, self-reflective adaptations of these very genres. The first case study examines the extent to which VR can remediate not only entertaining cinema and gaming but also the critical negotiation of authenticity in documentary film. Conversely, the second example, a Taiwanese remake of the Moon Landing, labeled as fake by conspiracy theorists, is discussed with regard to the possibilities of mockumentaries, political satire, and science in VR. According to Cavaletti, Conte, and Pinotti, VR reception is thus realized not in the overwhelming power of a problematic medium of empathy, but »in the complex interplay between narrative construction, user performativity, and awareness of the artifactual nature of the medium.«

In the last contribution, **Manischa Partowi** offers a unique perspective on the reception of VR art. In her article *Orchestrating Expectations, or Technoromanticism in VR Art*, she juxtaposes two contemporary artistic VR examples. As becomes clear from her close reading of Jordan Wolfson's *Little Room* and Theo Triantafyllidis' *Staphyloculus*, both works invest in creating an extreme situation by physically subjugating the user to the technology. While these extraordinary situations contradict any sense of normality, everydayness is brought back into focus through various aspects of the socially coded exhibition space. At this interface between the physical and the virtual space of the artwork, the author situates the orchestration of expectations, which she identifies as a fundamental artistic strategy for engaging with and reflecting a technoromantic imaginary of VR. In Wolfson's work, Partowi recognizes the use of VR's immediacy effect as a means of fetishizing the technical apparatus. In Triantafyllidis' work, on the other hand, she observes an exposure of VR's immediacy effect, revealing the technological complexity of bodily involvement within a virtual space.

Bibliography

Akbal, Zeynep (2023): Lived-Body Experiences in Virtual Reality: A Phenomenology of the Virtual Body, Bielefeld: transcript. DOI: https://doi.org/10.1515/9783839466766

Anthropy, Anna/Julia Stoschek Foundation (2024): Worldbuilding. Videospiele und Kunst im digitalen Zeitalter, Berlin: Hatje Cantz Verlag.

Aarseth, Espen/Stephan Günzel/Erik Hoops (2019): »Introduction. Space – The Theoretical Frontier«, in: Id. (eds.), Ludotopia, Bielefeld: transcript, pp. 7–10. DOI: https://doi.org/10.14361/9783839447307-001

Beil, Benjamin et al. (eds.) (2025): Gaming the Metaverse, Bielefeld: transcript.

Bieger, Laura (2015): Ästhetik der Immersion. Raum-Erleben zwischen Welt und Bild. Las Vegas, Washington und die White City, Bielefeld: transcript.

Bolter, Jay David/Richard Grusin (2000): Remediaton. Understanding New Media, Cambridge, MA: MIT Press.

Bonner, Marc (2018): »Gekerbte Wildnis – Inszenierungen vermeintlich unberührter Umwelt in digitalen Spielwelten«, 28.02.2018, in: Paidia: Zeitschrift für Computerspielforschung, online: https://paidia.de/die-gek

erbte-wildnis-inszenierungen-vermeintlich-unberuehrter-umwelt-in-digitalen-spielwelten/ (last access: 15.11.2025).

Blumenberg, Hans (2010): Theorie der Lebenswelt, Berlin: Suhrkamp.

Bunnenberg, Christian et al. (2025): »Von Empathie zum Caring. VR als soziale Konstellation«, in: Patrizia Breil/Florian Sprenger (eds.), Virtuelle Universität – Geistes- und gesellschaftswissenschaftliche Zugänge, Bielefeld: transcript, pp. 461–474.

Dekker, Annett (2021): »Curating Digital Art. From Presenting and Collecting Digital Art to Networked Co-Curation«, in: Id. (ed.), Curating Digital Art, Amsterdam: Valiz, pp. 14–33.

Eichwalder, Manischa (2024): »Embracing Otherness? A VR Body Hack by Morehshin Allahyari«, in: AN-ICON. Studies in Environmental Images 3/1, pp. 119–132. DOI: https://doi.org/10.54103/ai/23206

Ernst, Christoph/Regina Ring/Oliver Ruf/Jens Schröter/Sabine Wirth (eds.): Interface Studies. Eine Einführung, Tübingen: Narr (UTB) 2025 (forthcoming).

Fedorova, Ksenia: Tactics of Interfacing. Encoding Affect in Art and Technology, Cambridge, MA: MIT Press 2020.

Grau, Oliver (2004 [2001]): Virtual Art. From Illusion to Immersion, Cambridge, MA: MIT Press.

Günzel, Stephan: Egoshooter. Das Raumbild des Computerspiels. Frankfurt am Main/New York: Campus Verlag 2012.

Handout for the exhibition of DiMoDA 4.0: Dis/Location at Bannister Gallery, Rhode Island, 6.-28.10.2022.

Hattinger, Georg/Peter Weibel (eds.) (1990): Ars Electronica 1990. Digitale Träume – Virtuelle Welten, Linz: Veritas.

Himmelsbach, Sabine/Boris Magrini (eds.) (2023): Algorithmic Imaginary, Basel: Christoph Merian Verlag.

Kazarian, Choghakate/Camille Lévêque-Claudet (eds.) (2023): Immersion. The Origins: 1949–1969, Ex. Cat., Paris: Hazan.

Kornbluh, Anna: Immediacy, or The Style of Too-late Capitalism, London/New York: Verso 2023.

Kolesch, Doris (2021): »Ästhetik der Immersion«, in: Georg W. Bertram/Stefan Deines/Daniel Martin Feige (eds.), Die Kunst und die Künste. Ein Kompendium zur Kunsttheorie der Gegenwart, Berlin: Suhrkamp, pp. 422–441.

Krysa, Joasia (ed.) (2006): Curating Immateriality. The Work of the Curator in the Age of Networked Systems 3, New York: Autonomedia.

Kuni, Verena (2015): »F (ANALOGITAL)«, in: Daniel Kulle/Cornalia Lund// Oliver Schmidt/David Ziegenhagen (eds.), Post-digital Culture, http://www.post-digital-culture.org/kuni (last access 16.11.2025).

Kwastek, Katja: Aesthetics of Interaction in Digital Art, Cambridge, MA a. o.: MIT Press 2013.

Luckraft, Paul/Elizabeth Neilson (eds.) (2021): Artists in Virtual Reality, London: Zabludowicz Collection.

Magrini, Boris (2023): »Kunst und das Metaverse heute – ein Wettbewerbsfeld«, in: Pamela Scorzin (ed.), Mixed Realities. Neue Wirklichkeiten in der Kunst, Kunstforum International 290, pp. 67–75.

Milgram, Paul/Fumio Kishino (1994): Taxonomy of Mixed Reality Displays, in: IEICE Transactions on Information and Systems E77-D/12, pp. 1321–1329.

Nakamura, Lisa (2019): »Feeling Good About Feeling Bad: virtuous virtual reality and the automation of racial empathy«, in: Journal of Visual Culture 19/1, pp. 47–64.

Partowi, Manischa/Annette Urban/Manuel van der Veen (2025) (eds.): VR in der Gegenwartskunst. Ein Living Catalogue, online: www.vr-in-gegenwartskunst.de (last access: 15.08.2025).

Partowi, Manischa/Annette Urban/Manuel van der Veen (2025): »Screenshot«, in: Patrizia Breil/Florian Sprenger (eds.), Virtuelle Universität – Geistes- und gesellschaftswissenschaftliche Zugänge, Bielefeld: transcript, pp. 383–392

Reich, Julia/Manuel van der Veen (2023): »A kind of Mixed, Intermediate Experience. On the Entanglement of Image and Bodies«, in: Yearbook of Moving Image Studies (Mixed Reality Images), Darmstadt: Büchner.

Reich, Julia/Annette Urban/Manuel van der Veen (2023): »passthrough. Von Portalen, Durchblicken und Übergängen zwischen den (virtuellen) Welten«, in: Pamela Scorzin (ed.), Mixed Realities. Neue Wirklichkeiten in der Kunst, Kunstforum International 290, pp. 86–95.

Reinfeld, Philipp (2022): »Performatives Entwerfen in Virtual Reality: Körperaktive Hybridisierungen zwischen virtuellem und realem Raum«, in: Kassandra Nakas/Id. (eds.), Bildhafte Räume, begehbare Bilder, Leiden: Brill/Fink, pp. 27–51. DOI: https://doi.org/10.30965/9783846767238_004

Rieger, Stefan/Armin Schäfer/Anna Tuschling (eds.) (2021): Virtuelle Lebenswelten. Körper – Räume – Affekte, Berlin/Boston: De Gruyter.

Rupert-Kruse, Patrick (2024): »Panorama – Sphäre – Skulptur. Versuch einer Narratopologie virtueller Realitäten«, in: Lars C. Grabbe/Id./Norbert M. Schmitz (eds.), Neue Erzählformen in dynamischen Bildtechnologien.

Formprobleme zwischen Populärkommunikation und autonomer Kunst, Marburg: Büchner, pp. 83–107.

Sauerländer, Tina (2023): Unleashed Utopias. Artistic Speculations about Today and Tomorrow in the Metaverse, Ex. Cat., VR Art Prize Berlin, online: https://vrkunst.dkb-stiftung.de/export/sites/vrkunst/en/documents/UU_katalog_EN_final.pdf (last access: 10.10.2025).

Schröter, Jens (2004): Das Netz und die Virtuelle Realität. Zur Selbstprogrammierung der Gesellschaft durch die universelle Maschine, Bielefeld: transcript.

Scorzin, Pamela (2023): »The ›Phygital‹ as the Virtual Real. The Role of Mixed Realities in Contemporary Art«, in: Yearbook of Moving Image Studies (Mixed Reality Images), Darmstadt: Büchner, pp. 115–135.

Tuin, Iris van der/Nanna Verhoeff (2022): Critical Concepts for Creative Humanities, Lanham a. o.: Rowman & Littlefield 2022.

Urban, Annette (2025): »Über Portale, Sphären und andere eigensinnige Objekte als Weltbezüge im künstlerischen Worldbuilding mit Mixed Realities«, in: Patrizia Breil/Alisa Kronberger (eds.), Eigensinnige Objekte. Virtuelle Möglichkeitsräume zwischen Aufforderung und Entzug, Bielefeld: transcript, pp. 101–141.

Urban, Annette (2023): »Daten, Atmosphären, Texte – Künstlerische Erfahrungsräume und Virtualisierungen des Literarischen«, in: Journal of Literary Theory 17/1, pp. 111–142. DOI: https://doi.org/10.1515/jlt-2023-2006

Urban, Annette (2023): »Escape. Weltbezug und Weltentzug in Kunst mit VR«, in: Lilian Haberer/Karina Nimmerfall (eds.), Mouvement/Movement. Festschrift für Ursula Frohne, Munich: edition Metzel, pp. 99–104.

Urban, Annette (2026): »Insuläre Topografien und welthafte Ganzheit. Künstlerische Auseinandersetzung mit Landschaft in VR«, in: Anna Polze/Manuel van der Veen (eds.), Virtuelle Landschaften. Raumerkundungen an der Grenze des Screens, Bielefeld: transcript (forthcoming).

van der Veen, Manuel (2025): »Headset – Maske«, in: Zeitschrift für archäologische Aufklärung 4, Bielefeld: transcript.

van der Veen, Manuel (2025): »Objekte«/»Second Screen«, in: Manischa Partowi/Annette Urban/Manuel van der Veen (2025) (eds.): VR in der Gegenwartskunst. Ein Living Catalogue, online: www.vr-in-gegenwartskunst.de (last access: 15.08.2025).

Image Credits

fig. 1: © Courtesy the artist, photo: Alessandro de Matteis
fig. 2: © Courtesy the artist, photo: J. Pegman
fig. 3: © Courtesy the artist, source: hairygaze
fig. 4: © DiMoDA
fig. 5: © DiMoDA

ART/GAMES

Genealogies of VR

The Afterlife of Antiquity in Virtual Reality

Ursula Ströbele

Introduction: Floating Sculptures

In their film *The 3D Additivist Manifesto* (2015; 10:11 min), Morehshin Allahyari and Daniel Rourke proclaim a new posthumanistic age. »We call not for passive, dead technologies,« a voiceover intones, »but rather for a gradual awakening of matter, the emergence, ultimately, of a new form of life.«[1] The film positions the 3D printer as a profound metaphor, a technology for channeling creative endeavor through digital processes. Allahyari and Rourke's manifesto aims for a cybernetic interconnection of machine and human body. *The 3D Additivist Manifesto* (fig. 1) teems with sculptural quotations, positing a significant reconnection to the history of canonized sculpture, ranging from the iconic readymade *Fountain* (1917), representing modern notions of art and an expanded concept of authorship, to the Hellenistic torso of the *Venus de Milo*, enshrined in the traditional aesthetics of the Western sculpture canon as a feminine ideal. Departing from linear concepts in history, here, digitized sculptural objects, industrial artifacts, technological residues, cyber bodies, and animalistic details, such as the octopus, all simultaneously coexist. They float in the stream of oil from which they have been printed. This elemental substance, forged over millions of years, has enabled innovative technologies

1 Moreshin Allahyari/Daniel Rourke (2015): »The 3D Additivist Manifesto«, in: additivism.org. Online: https://additivism.org/manifesto/ (last access: 16.08.2025). See Ursula Ströbele (2023): »Sculpting Digital Realities. Notes on Truth to Materials, the Aesthetic Limit, Site-Specificity and 3D-Printing«, in: Mara-Johanna Kölmel/Id. (eds.), The Sculptural in the (Post-) Digital Age, Berlin/Boston: Walter de Gruyter, pp. 83–101, here: pp. 99–100. Thank you to Manischa Partowi, Annette Urban, and Manuel van der Veen for inviting me to the workshop and publication *Virtual Reality Exhibited. Interfacing Art, Games and Everyday Life* at Ruhr University Bochum and for the inspiring exchange.

such as 3D printing – a process that itself evokes deep time. Here, synthetic technology becomes a new modality of a biological posthuman medium, attributing intelligence to all kinds of matter. The *3D Additivist Manifesto* demonstrates the multiplicity inherent to the expansion of the sculptural, which entails an understanding of materiality that transcends the antique dualism of form and content and incorporates a relational field.[2]

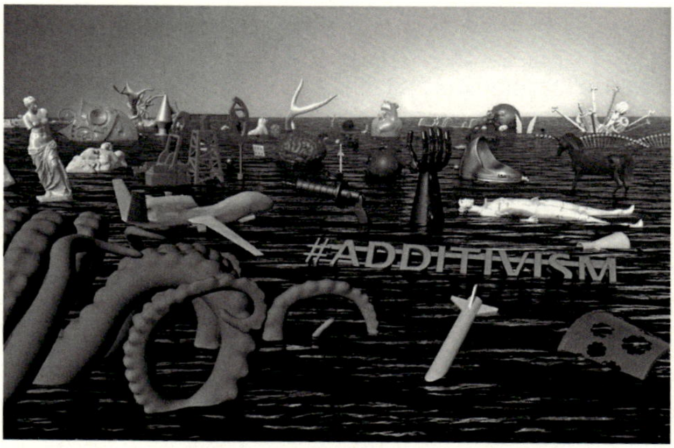

fig. 1: Morehshin Allahyari & Daniel Rourke: The 3D Additivist Manifesto, 2015, 10:11 min, sound design by Andre Young, https://additivism.org/manifesto/

1. Quoting the Canon in Virtual Reality Art: *Mercury*, the Antique Messenger of the Gods Traveling Between the Worlds

Based on the *3D Additivist Manifesto*, I would like to present a few thoughts on how digital art–especially Virtual Reality (VR) works – contains sculptural quotations or conjures an antiquity that never existed in the virtual world. Which

2 Cf. M. Allahyari/D. Rourke: The 3D Additivist Manifesto. »Additivism« is derived from »additive« and »activism« and, according to the artists, aims to disrupt existing categories, expanding the art project toward an online community, activism, ironic commentary, and revolutionary potential.

objects are cited for this genealogical inscription within an (as yet) still-forming canon? How are these so-called antique set pieces, or *Versatzstücke*, situated between adaptation and alienation, and what role do speculative archaeology and humor play? And is »quotation« even a suitable term to describe these diverse modifications? The term quotation – an essential part of academic work – derives from the Latin *citatum*, meaning »to lead« or »to summon,« and from the verb *citare* »to set in motion,« as in citing someone in court.[3] It is the substantivized second participle of *citare* and traditionally refers to a passage taken verbatim or in terms of content from a text, intended to acknowledge the original author or copyright. Content can also be taken from other media, such as image or sound quotations. My text addresses quotations from sculpture and architecture, or their speculative references, which literally »set in motion« new ways of dealing with semantically overloaded cultural heritage. Notions of antiquity, canon, and material are at the center of the following considerations.

VR works enable an immersive sensory engagement of the recipients, who »wear« part of the apparatus, equipped with a head-mounted display, enter the digital spatial image and navigate through a virtual space of experience.[4] Among the early VR experiences is Banz & Bowinkel's *Mercury* (2016) (fig. 2), created nearly ten years ago for the HTC Vive. Wearing goggles, users stride through a virtual setting resembling an archipelago with five circular islands connected by narrow bridges, accessible by using the controller. They hover Olympus-like over a mountainous landscape close to the sea, which may cause dizziness or a touch of vertigo in some people while moving around. The platforms feature natural elements such as birch trees, skeletal-looking conifers, and a few rocks, alongside constructed forms: an architectural cube seemingly made of concrete, a hemispherical pavilion with a grid-like triangular structure, enlarged screens of a digital device, and text-based signs – such as »shift,« »function,« and »control« – that reference communication and the entanglement of real life and virtual worlds. Huge shiny spheres mirror an arena, bouncing up and down. Different avatars, some with fragmented mannequin-

3 See e.g. Digitales Wörterbuch der deutschen Sprache (DWDS), in: dwds.de. Online: https://www.dwds.de/wb/Zitat (last access: 18.08.2025).

4 Banz & Bowinkel (2016): »Mercury (VR)«, in: banzbowinkel.de. Online: https://www.banzbowinkel.de/project/mercury/ (last access: 16.08.2025). For the history of virtual reality see e.g. Oliver Grau (2001): Virtuelle Kunst in Geschichte und Gegenwart. Visuelle Strategien, Berlin: Reimer; Sabine Himmelsbach (2021): »Unbound Spaces – Immersive Virtualities«, in: Tina Sauerländer (ed.), Resonanz der Realitäten, VR Art Price of DKB in Cooperation with CAA Berlin, Ex. Cat., Haus am Lützowplatz, Berlin, pp. 20–25.

like bodies, appear repeatedly. Ghostly and prancing through the scenery, they seem absorbed in their own movements as they perform slow dance gestures, circling through the space. The lighting situation slowly changes as you walk, getting darker at the end of the »day« before the brightness increases again. These virtual entities are all intertwined into a surreal scenery aesthetically borrowed from gaming worlds, in which known physical laws have become fluid, without being a direct quotation such as in the following case. In the center of the artwork, but also of my analysis, is an ancient statue: Mercury, the winged messenger of the gods. Four avatars dressed in black and pink tights surround him, moving similarly to the mannequin-like figures mentioned above. Another version of the antique quote is a shattered statue, located next to the dry conifers, whose splinters can be virtually walked upon. User may even pass through the virtual stone – an indication of digital materiality and its permeability.

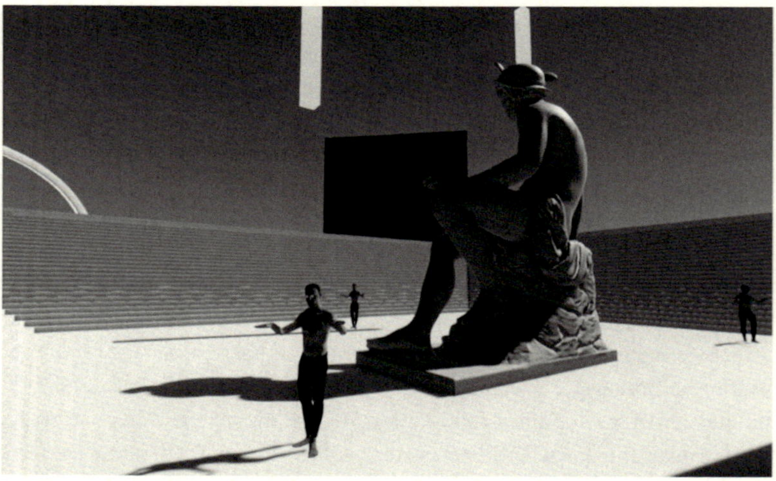

fig. 2: Banz & Bowinkel: Mercury, 2016-2017, interactive virtual reality installation for HTC Vive, customized computer, head-mounted display, 3D-printed porcelain interface button, web cameras, monitor, plexiglass, various cables; set up in a black anodized aluminum frame structure

2. Marble in the Virtual World: Material Simulation/Material Knowledge

In addition to the two aforementioned citations from *Mercury*, the second work by Banz & Bowinkel, created a year later, is evidence of a further development of the ancient statue. In *Palo Alto* (2017) the digital sculpture is animated.[5] It shatters and reassembles itself (fig. 3), while maintaining the illusion of a compact, three-dimensional form. Furthermore, Mercury changes his material properties. While at one moment the statue seems to be made of elastic, wobbling rubber, swinging up and down, the next moment it mutates into a figurative block of hard marble again or bursts into fragments. The users only have an indirect influence on the changing aggregate state. If one finds the entrance to the main tower, they reach the next level, which has different colors and lighting. The statue changes as soon as you enter or leave this passage, thus altering your own »gaming« level within this responsive environment. The aesthetic boundary between the sphere of art and the viewer's space, as discussed by Ernst Michalski (1931), seems to have disappeared or at least been reduced.[6] Only eruptive image transitions and glitches can disturb the spatial illusion and immersive effect. Oliver Grau underlines the reduction of the aesthetic distance and its manipulative effect:

> »The more ›natural‹ the interface, the more pronounced not only the danger that the invisible part of the ›technological iceberg‹ remains closed and unconscious to its user, but above all the more intense the illusionary dissociation with the data space.«[7]

---

5 Banz & Bowinkel (2017): »Palo Alto (VR)«, in: banzbowinkel.de. Online: https://www.banzbowinkel.de/project/palo-alto/ (last access: 16.08.2025). The work was amongst others exhibited in *The Unframed World: Virtual Reality as artistic medium* at House of Electronic Arts (HEK) Basel in 2017, curated by Tina Sauerländer, in *Brave New Worlds. Virtual Realities in Contemporary Art* at Zeppelin Museum in 2017/18, curated by Claudia Emmert and Ina Neddermayer, at Kunstverein Arnsberg 2019 in the exhibition *In Medias Res*, curated by the author. It is now part of the permanent collection of the Museum Kunstpalast in Düsseldorf.

6 Ernst Michalski (1996): Die Bedeutung der ästhetischen Grenze für die Methode der Kunstgeschichte, Berlin: Gebr. Mann Verlag.

7 O. Grau: Virtuelle Kunst in Geschichte und Gegenwart, p. 183 (translation by the author). He does not mention Ernst Michalski.

The so-called material time and the (learned) material knowledge itself, which includes the specific material properties, generates a dynamic tension between classical object aesthetics, sculptural *memoria*, and the digital counterfeit.[8] As is well known, marble, respectively stone, was celebrated as being timeless due to its durability. Here, it mutates into a digital simulation that is programmed on the computer and consists of individual pixels, as well as the texture map and normal map, which have an even greater influence on the artistic work on digital materiality.

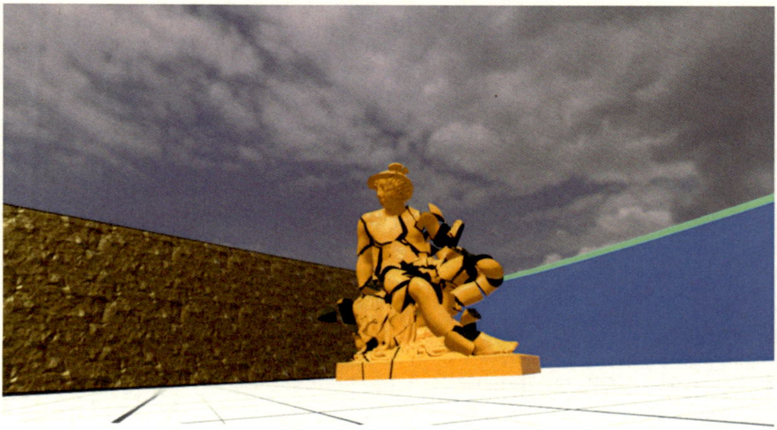

fig. 3: Banz & Bowinkel: Palo Alto, 2017-2018, interactive virtual reality installation for HTC Vive, customized computer, head-mounted display, monitor, carpet, various cables; set up in a four-colour powder coated anodized aluminum frame structure

On one hand, the alteration of material properties and the visual alienation of classical references carries a humorous component and could even be interpreted as a temporary iconoclastic act, which, however, remains without consequences due to its virtuality and is reversed again by the artistically programmed code. On the other hand it might allude to the elasticity of histori-

8 For more information on the material turn, see the recent insightful summary by Beate Fricke/Ann-Sophie Lehmann (2024): »Materials Matter«, in: Kunstchronik. Monatsschrift für Kunstwissenschaft 77/7, pp. 461–467, https://doi.org/10.11588/kc.2024.7.105329.

cal linearity, stretching the narrative of (progressive) history within the virtual world. There is also an iconoclastic moment in the splintering or crumbling of *Mercury*. The alienation or transformation of the quotation reveals the medium itself, similar to glitches, while simultaneously exposing the playful possibilities of digital art. Thus, a reference to antiquity in the form of an iconic statue is virtually cited, »herbeizitiert« (summoned) as a set piece and uprooted fragment of a larger cultural-historical context. The rubbery messenger of the gods looks like a humorous estrangement, reminiscent of merchandising products and miniaturized souvenirs, yet he is actually monumentalized in relation to the architectural surroundings and avatars. Does VR function here as a technology of counter-history and estrangement or does it imply a certain conservatism? Echoing the Eurocentric canon, the work reflects both an attempt at self-inscription within an art-historical genealogy and a humorously critical distancing qua alienation and symbolic destruction of historical continuity. While the Dadaists celebrated the elasticity, flexibility, adaptability, and ubiquity of industrial rubber as an artistic principle of life and survival in their collages and assemblages, Banz & Bowinkel playfully explore the facets of digital material illusion in this work.[9] Often, playful aspects have something childlike, degrading, and a closeness to life about it, which classical art history attempted to separate from so-called high art, whereas it is a decisive characteristic of virtual art.

The popularity of antique references, or spolia in the broader sense, is also evident in Robin Rutenberg's VR-work *Stonesong* (2022) (fig. 4): a pair of antique columns stand as isolated objects framing stairs without any corresponding building, next to a free-floating marble torso.[10] A few seconds later, the viewer experiences overgrown columns as if deep time had undergone a paradoxical acceleration; the figures look like petrified, stone-skinned fossils in an Arcadian setting.[11] They explain: »When journeying through this ethereal

9 Ursula Ströbele (2024): »Elastikakt – Elastizität als künstlerisches (Über-)Lebensprinzip«, in: Julia Wallner (ed.), Der Die Dada. Unordnung der Geschlechter, Ex. Cat., Munich: Hirmer, pp. 38–45.

10 They initially presented the work as a single VR headset installation with headphones. According to the artist, a video out ran to a large 55" flatscreen TV and an audio out fed two additional pairs of headphones, while the VR participant's journey could be viewed and listened to by other recipients. The statues remain stationary, while the stones are weightless, shapeshifting.

11 A summary can be found here: Robin (27.10.2022), Vimeo: https://vimeo.com/764574164 (last access: 16.08.2025).

realm, the listener encounters multiple 3D rendered statues of the artist and other stone structures forged from transcendent matter,« referring, amongst others, to the singing *Colossi of Memnon*, a pair of twin Egyptian burial monuments constructed by Pharaoh Amenhotep III. Rutenberg inscribes themself via their own portrait in the aesthetic genealogy of an ancient statue. Statues, or rather their shells, change color, pulse with light or are enveloped in mist. They »[swell] and [surge],« spilling »over [their] own bodily boundaries,« performing an allegorical image for materiality and its becoming.[12] Here again, this includes a reference to stone as a fundamental artistic material, which could not be more contrary to the (seeming) immateriality of VR/digital art. But in contrast to physical sculptures, it is the skin of a non-monolithic torso with fluid, porous contours that drifts apart, splinters, and reassembles in small particles – similar to *Mercury*. The porous shell, its deceptive appearance and false façade, refers to the timeless canon of antique statues.

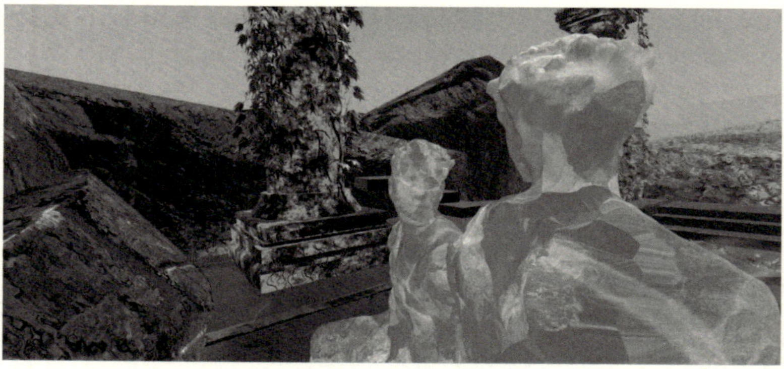

fig. 4: Robin Rutenberg: Stonesong, 2022, VR poem and soundscape installation

Even if new modes of perception are made possible, only the »visual aid« of the interface reveals the sculptural work. Despite the dissociation between the practically non-existent movement of the viewer's physical body and their

12 Robin Rutenberg (2022): Stonesong. A VR Soundscape and Poem, Sound Studies and Sonic Arts Master Exhibition, Berlin University of the Arts, Collegium Hungaricum Berlin, booklet. Thank you to Tina Sauerländer, who draw my attention to this project.

movement in cyberspace, the apparatus remains palpable. This dynamic coupling of body and virtual space, as well as the indiscernibility of perception and affection, is an important aesthetic feature of VR. Our own knowledge, experience, and memory of material properties, such as the stability and rigidity of marble (*Mercury*) or the rough surface of wood/bark, still determine our perception of phenomena and sculptures in the virtual world and also determine how they affect us physically, despite being aware of the digital illusion.[13] Henri Focillon describes this as »tactile faculty,« which only allows »the possession of the world«[14]: »The hand knows that objects are weighted, that they are smooth or rugged, that they are not fused to the back of the sky or the earth like they appear.«[15]

3. Artistic Strategies of Reproduction and Appropriation

The canonical antique statue of Mercury, a central reference for both works by Banz & Bowinkel, is a quote within a quote within a quote: like ancient statues such as *Hermes Fastening His Sandal* (2nd c. AD), the French sculptor Jean-Baptiste Pigalle conceived his seated *Mercury* at the moment he fastens his heel-wings, preparing to take to the skies on an errand. The smaller-than-life marble sculpture was submitted in 1744 as Pigalle's reception piece for admission to the Royal Academy in Paris. By the 17th and 18th centuries, the imitation of antiquity no longer functioned as a form of slavish copying: *imiter* is now clearly separated from *copier*, as the French court historian and co-founder of the Paris Academy André Félibien claims. »When one speaks of imitating antiquity or the style of a specific master,« he writes in *De principes,* »this does not mean to copy every drawn, painted or sculpted detail, but rather to formulate a similar idea and work in the same style.«[16] This definition of *imiter* underscores how

13 Technological accessibility and implicit dependency include the respective (visual) equipment, i.e., software and hardware, which are equally essential for production and reception, require an audience that owns a digital device and has internet access, possibly access to VR glasses, and at the same time reflect media history.

14 Henri Focillon (2018): In Praise of Hands, New York: Parkstone Press International, p. 28.

15 Ibid.

16 André Félibien (1699): Des principes de l'architecture, de la sculpture et de la peinture et des autres arts qui en dépendent: avec un dictionnaire des termes propres à chacun de ces arts, Paris (Vol. 3), p. 442 (translation by the author).

it is not the exact replication of the ideal sculpture that matters, but the study of nature, anatomy, and its processes – and the subsequent reinterpretation of these elements. Since then, there have been various adaptations of the messenger of the gods in different attitudes and temporal moments. In an interview, Banz & Bowinkel explain their personal motivation:

> »The only direct reference is a 3D scan of a statue of Mercury from a project by Oliver Laric. We are interested in the figure of Mercury as the ancient god of commerce as well as thievery. An apt symbol of our times. We also wanted to implement the project instigated by Laric.«[17]

This echoes an afterlife of the Eurocentric ancient sculptural canon.[18] In recent years, Oliver Laric has scanned several antique, Greco-Roman statues, including *Mercury in Repose* carved by Joseph Nollekens in the 18th century. The file can be downloaded from a specific web portal, such as Turbosquid, and 3D printed.[19] Laric's sculptures are also not straightforward copies of their models. His work *Sleeping Boy* (2016) is printed in different materials. Such sculptures remind us of earlier strategies, for example in the Renaissance, when artists completed Greek and Roman torsos in marble (or in drawings) and often interpreted them self-consciously, for example following the excavation and discovery of the Laocoon group in 1506. In the context of Laric's engagement with the legacy of antique models, Melissa Gustin thus speaks of »digital neo-

---

17 Banz & Bowinkel in an interview with Dominik Busch (2017), in: Ina Neddermeyer/Claudia Emmert (eds.), Brave New Worlds. Virtual Realities in Contemporary Art, Ex. Cat., Zeppelin Museum, Friedrichshafen, p. 59.

18 For the notion afterlife, see also Anton Springer (1886): »Das Nachleben der Antike im Mittelalter«, in: Id. (ed.), Bilder aus der neueren Kunstgeschichte (Band 1), Bonn (2nd, revised and expanded edition). In his chapter »The Afterlife of Antiquity« he elaborates on the world of ancient Greece and Rome and how it has inspired the Western cultural history, how it has left many traces and has been a reference, still until today, one could add. Throughout his life, Aby Warburg devoted himself to antiquity and its afterlife in images, such as in the introduction to the *Mnemosyne Atlas*, where he formulates the idea of the long-term cultural impact of images (Aby Warburg (2012): »Der Bilderatlas Mnemosyne«, in: Gesammelte Schriften. Vol. II/1. Ed. by Martin Warnke/Claudia Brink, Berlin: Akademie Verlag). Thanks are due here to Steffen Haug.

19 Bridget O'Neal (2014): »Artist Oliver Laric Truly ›Shares‹ his Lincoln 3D Scans from Usher Gallery Exhibit«, in: 3dprint.com. Online: https://3dprint.com/16512/oliver-laric-lincoln-3d-scans/ (last access: 16.08.2025).

classicism,« situating his artistic practice within the history of neoclassical imitation and legacy of antique models.[20]

A further iconographic layer arises from the various roles embodied by Mercury, a god of financial gain, commerce, eloquence, communication, travelers, and thieves who also guides souls to the underworld and serves as the »messenger of the gods«. Today, big data and the global, often free circulation of digital commodities lead to a dilemma that is linked to the question of how we define and deal with value in our society, leading Banz & Bowinkel to ask: »Is this trade or theft? Or, who profits from a certain structure of added value and who does not?«[21] Using digital editing and remixing, one might ask, who quotes whom and why? A quotation can also become intellectual theft if it is not properly identified. As Dennis Rudolph provocatively explains in *The Portal* (2022): »I stole all these figures from a huge baroque ceiling painting by Cortona in Rome. I started repainting them in 3D in the VR. Kind of updating them.«[22] The artist refers to his series of giant 3D painted mythological figures, which embody a painterly digital plasticity. His VR experience reshapes Auguste Rodin's *Gates of Hell* (1880-circa 1890) – another art historical quotation.

A further example of an antique reference is the arena-like architecture reminiscent of the antique agora, which is reflected in the bouncing spheres and appears miniaturized and distorted on their surfaces. *Palo Alto* is also determined by set pieces of reality, such as a virtual stonewall, which is juxtaposed with a blue concave-shaped mural element. Its color refers to the aesthetics of the blue screen and does not represent any architectural detail; rather, it embodies the omnipresence of the internet.[23] Scattered deciduous trees stand on a grid-like floor; avatars dressed in black and pink overalls move

---

20 Melissa L. Gustin (2024): »Do Sleeping Shepherds Dream of 3D-Printed Sheep: John Gibson, Oliver Laric, and Digital Neoclassicism«, in: British Art Studies, London/New Haven: Paul Mellon Centre for Studies in British Art and Yale Center for British Art, https://britishartstudies-24.netlify.app/digital-neoclassicism/ (last access: 16.08.2025).

21 Banz & Bowinkel in an interview with Dominik Busch, Brave New Worlds, p. 59.

22 Dennis Rudolph (2024): »The Portal (2024). VR Experience for Meta Quest«, in: radiancevr.co. Online: https://www.radiancevr.co/artists/dennis-rudolph/, 9:21–9:31 min (last access: 16.08.2025).

23 The title, *Palo Alto*, refers to Silicon Valley, which, quoting the artists, »symbolizes the cradle of the digital revolution.« (Banz & Bowinkel in an interview with Dominik Busch, Brave New Worlds, p. 59).

through the scene; light monochrome clones fall from a tower, in whose roof truss a pendulum swings back and forth between »true« and »false«; the text on the terrain here reads: »A place where none« and »See for be seen« – quotes from *Worstward Ho* (1983) by Samuel Beckett.

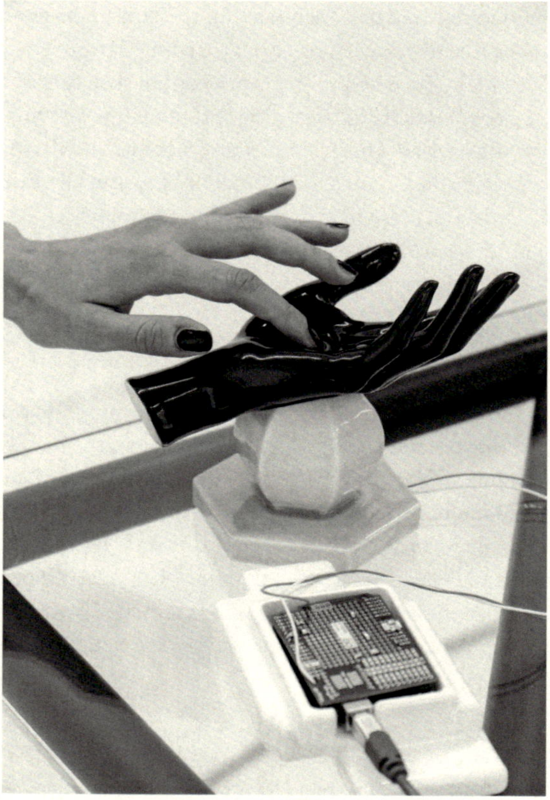

fig. 5: Banz & Bowinkel: Mercury, 2016–2017, interactive virtual reality installation for HTC Vive, customized computer, head-mounted display, 3D-printed porcelain interface button, web cameras, monitor, plexiglass, various cables; set up in a black anodized aluminum frame structure, installation view: Substance, DAM Gallery, 2017

Similar to Mercury, who continuously travels between realms, Banz & Bowinkel's work unites the two worlds: in VR, one of the screens displays the exhibition space along with the hands of the viewer and both controllers, while the other screen shows analogously an outside perspective onto the users. With a webcam it is clamped in an open aluminum frame that follows the changes of the virtual landscape, enabling other viewers to perceive the virtual movements and gain insight into the programmed scenery even without wearing the head-mounted display. The Italian philosopher Luciano Floridi (2015) names this contemporary entanglement of our post-digital world »onlife.«[24] The computer handling large amounts of data is installed as an aesthetic object next to the webcam, as is a 3D printed porcelain hand with a button (fig. 5), placed on a ball, suggesting a connection to the physicality of »traditional« sculpture. The button in the porcelain hand can be used to accelerate the day-night cycle in the VR world. The hand appears here across multiple registers: the user's hands holding the controller; as a sculptural object in the exhibition space, referring to interfaces that have become an extended body part and multifunctional prosthesis; as a digital counterpart in VR, and indirectly through the visible and tangible controller; and finally, as the programming hands of the artist duo. From an art historical perspective, the artist's hand is connected to various topoi; as a tactile organ, it represents the touch and modeling of material and is considered the executive force of the creative artistic genius. Henri Focillon praises them in his poetic odes to the hand (1934) as »tireless companions.« They are responsible for making man become »aware of the difficulty of thought [...] the hand is about action: it takes, it creates, and it could be said that it thinks.«[25] Even though vision is often regarded as the central organ for the perception of aesthetic experience in digital art, the sense of touch also plays a decisive role. Derrick de Kerckhove has outlined the sense of touch as an essential sense of interactive media and each user's proprioception. His examples of scanning the television screen with the eyes, and of the tactile functions of the remote control and video recorder, could be expanded to consider these implements as historical precursors to the controllers and head-mounted display used in VR.[26] Shirin

24 Luciano Floridi quoted after S. Himmelsbach: Unbound Spaces, p. 21.

25 H. Focillon: In Praise of Hands, p. 11.

26 Derrick de Kerckhove (1993): »Touch versus Vision: Ästhetik neuer Technologien«, in: Wolfgang Welsch (ed.), Die Aktualität des Ästhetischen, Munich: Wilhelm Fink, pp. 137–168, here: pp. 145–148. See also Mark W. D. Paterson (2005): »Digital Touch«, in:

Weigelt also describes »tactility« as a central sensual-aesthetic category of digital media cultures.[27] Whether as a swipe, click, drag, or scroll gesture to browse and open digital pages, zoom in on individual details, or move with an avatar through a virtual landscape, navigating through a virtual world today is characterized by various haptic moments that even children can now easily master.

In *Palo Alto*, the display has slightly changed: a round, black carpet with the word »escape« rendered in neon green lettering and a photo-print depicting two avatars are added to the visible computer and the screen. The work can be experienced in different formats, whether as a site-adapted sculptural installation presented in an exhibition or collection context – such as *Palo Alto* in the Museum Kunstpalast in Düsseldorf – or a trailer-like excerpt on YouTube. The work itself can be experienced with Meta Quest goggles via the Radiance app at home or another chosen place with a WiFi-connection.[28]

4. Speculative Archeology and Scale in Digital Sculpture

The messenger of the gods cited by Banz & Bowinkel appears in a monumental format in relation to the avatars walking around it. There is an oversized, seemingly monstrous beetle – a quotation from Franz Kafka's *The Metamorphosis* (1915) – who also embodies the play with scaling. When you meet the insect with your avatar, your own size seems to change. In this virtual landscape, there are (no longer) any fixed distances or proportions for orientation. »The size is nothing; what matters is the scale,« Barnett Newman said in 1969.[29]

Constance Classen (ed.), The Book of Touch, Oxford/New York: Berg, pp. 431–436. His research has been dedicated to the sense of touch and haptic perception.

27 Shirin Weigelt (2019): »Tasten: Taktilität als Paradigma des Digitalen«, in: Rainer Mühlhoff/Anja Breljak/Jan Slaby (eds.), Affekt Macht Netz. Auf dem Weg zu einer Sozialtheorie der Digitalen Gesellschaft, Bielefeld: transcript, pp. 107–128, here: p. 155, https://doi.org/10.25969/mediarep/13223.

28 See Banz & Bowinkel (23.09.2017), Youtube: https://www.youtube.com/watch?v=Akp7cUB2bfQ; Banz & Bowinkel (21.02.2019, 2:22), Youtube: https://www.youtube.com/watch?v=flci8C6r8NY&t=142s; Banz & Bowinkel, in: radiancevr.co. Online: https://www.radiancevr.co/artists/banz-bowinkel/ (last access: 16.08.2025).

29 Barnett Newman quoted after James Meyer (2004): »No more Scale. The Experience of Size in Contemporary Sculpture«, in: Artforum 42/10, online: https://www.artforum.com/features/no-more-scale-the-experience-of-size-in-contemporary-sculptur

Yet while the physical body is still considered a reference in Abstract Expressionism and (Post)-Minimal Art, this orientation holds little relevance for digital art, which allows permanent (re)-scaling and floating, thereby provoking a kind of *sitelessness*.[30] But in VR and Augmented Reality (AR) digital devices, such as googles, controllers and smartphones or tablets themselves become forms of bodily prostheses and sites of localization and reference. Although we do not have visible avatar bodies as we walk through the virtual landscapes of *Mercury* and *Palo Alto*, we position ourselves in relation to the size of the virtual mannequins and clones, which makes the beetle, for example, appear monumental to us.

Size can be marshaled as a critique of traditional monumentality and of the portable commodity. Thus, according to David J. Getsy, »even if the statue is monumental or miniscule, the bodily sense of scale becomes a corporeal link between the viewer and the actual presence of the three-dimensional image [...].«[31] In VR, the users slip into their digital (physical) surrogate while teleporting through the virtual landscape. The corporeal link thus unfolds between each user's digital body, their individual memories of the actual, physical statue, and its virtual simulation, together suggesting a so-called speculative scale.

Speculation comes from the Latin *speculari*, meaning »to observe.« It describes a hypothetical train of thought that extends beyond tangible reality. »Speculation,« in the sense of Rosalyn Diprose, »is futural: it keeps open (past and present) worlds to potentiality, possibility, and the unknown«.[32] Potentiality is closely linked to virtuality; both terms embody a certain form of potentiality. Etymologically, virtuality is borrowed from the Latin *virtus* (»virtue,« »bravery,« »effectiveness«). In French, *virtuel* means »capable of acting,«, »existing as a possibility in accordance with its disposition.«[33] Virtuality is therefore an

---

e-168860/ (last access: 17.08.2025). See e.g. *Scale as Content*, Corcoran Gallery of Art, Washington DC, 1967.

30 See U. Ströbele: Sculpting Digital Realities, p. 100.

31 David J. Getsy (2014): »Acts of Stillness: Statues, Performativity, and Passive Resistance«, in: Criticism 56/1, pp. 1–20, here: p. 2.

32 Rosalyn Diprose (2017): »Speculative research, temporality and politics«, in: Alex Wilkie/Martin Savransky/Marsha Rosengarten (eds.), Speculative Research. The Lure of Possible Futures, London/New York: Routledge, pp. 39–51, here: p. 40.

33 See Sven K. Knebel (2001): »Virtualität«, in: Joachim Ritter et al. (eds.), Historisches Wörterbuch der Philosophie 11, Basel: Schwabe, pp. 1062–1066; Ralf Grötker (2001): »Virtuelle Realität«, in: Ibid., pp. 1066–1068 (translation by the author).

entity that is not physical but exists in its functionality or effect. The conceptual history of the virtual, from a potential force striving for actualization (Thomas Aquinas) and an element preceding reality in time (Henri Bergson) to the temporal latency and recursion of virtual images, would require a separate investigation.[34] Speculating, according to Noam Gramlich, is a linguistic-material assemblage with, through, and *in* bodies.[35] Gramlich suggests that »long before the speculative turn, thinking in *futurum II* and the visionary design of other pasts, presents, and futures were constitutive components of feminist, post-, and decolonial theories as well as gender and queer studies.«[36] Speculation – the term stands here as a counter-program to established and objective knowledge – considers the virtual reality-constituting power and significance of fictions and narratives. As an ontological endeavor, speculative strategies characterize both Banz & Bowinkel's *Mercury* and *Palo Alto*. This also applies to Juan Covelli's VR-work *Speculating the Fragmented Body* (2019) (fig. 6), in which he uses 3D scans of 14 Mesoamerican artifacts from the Stavenhagen Collection in Mexico City to create an online free access archive, motivated by his »[fascination] with the Zapoteca culture and its aesthetics.«[37] Covelli invited other artists to download and modify the objects, both to restore their visibility against oblivion and to grant these 2000-year-old archaeological objects a new digital life within a virtual gallery.[38] Rather than sculptural quotation, he is instead interested in remixing and reprinting these objects, altering them through code and creating glitches: »This act of reconfiguring and reinterpret-

34 See e.g. Eva Wilson (2016): »Hinter den Spiegeln. Virtualität, Rekursion und virtuelle Bilder im 19. Jahrhundert«, in: Heide Barrenechea/Marcel Finke/Moritz Schumm (eds.), Periphere Visionen. Wissen an den Rändern von Fotografie und Film, Paderborn: Wilhelm Fink, pp. 97–112.

35 Naomie Gramlich (2020): »Feministisches Spekulieren: Einigen Pfaden folgen«, in: Marie-Luise Angerer/Naomie Gramlich (eds.), Feministisches Spekulieren: Genealogien, Narrationen, Zeitlichkeiten, Berlin: Kulturverlag Kadmos, pp. 9–32, here: p. 21.

36 Ibid., p. 9.

37 Juan Covelli in an interview with Schloss-Post (2018): »Ancient Gods and Digital Power«, in: Digital Culture, Theory & Art 0, online: https://schloss-post.com/ancient-gods-digital-power/ (last access: 18.08.2025).

38 According to Covelli, »the artists who interpreted the works remained anonymous. [...] I was more interested in the act of reshaping and copying than in authorship. [...] challenging the notion of uniqueness. As a result, the work itself plays with a blurred concept of heritage and ownership.« (Juan Covelli in an e-mail to the author, 29.01.2025).

ing is a way of liberating the artifacts from their historical baggage and thus overcoming the colonial yoke,« he explains.[39] Covelli further argues:

> »The most effective tool we have in the digital domain is the capacity to copy and disseminate information endlessly; the power of the copy is important because it makes it impossible to control or to grasp an artifact. There is not just one digital artifact; there are infinite copies circulating the web in many forms. [...] We have passed from an era of mechanical reproduction to the era of the digital rhizome, a digital file that can be copied eternally, and can be viewed simultaneously in many places, a digital object thus has these magical properties of being ubiquitous, and this is what makes these new artifacts so powerful.«[40]

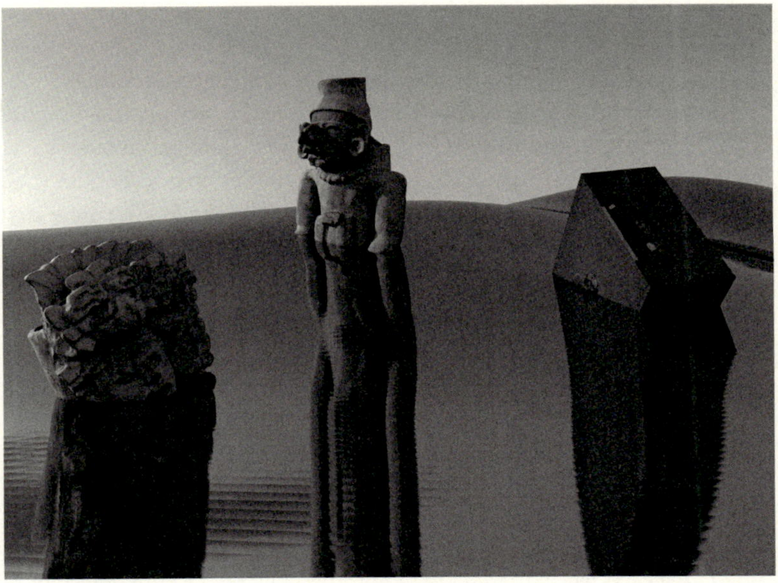

fig. 6: Juan Covelli, Speculating the Fragmented Copy, 2018

39 Web Residencies by Solitude and ZKM, online: https://speculating-the-fragmented-copy.schloss-post.com/index.html (last access: 18.08.2025).

40 Ibid.

Referring to their potential for sitelessness, these speculative sculptures were meant to be sent back to the artist for exhibition in a virtual gallery he created with the program Unity, which could be experienced through VR goggles. The gallery itself resembles a desert, »a future where the digital artifacts are the only thing remaining.«[41] Covelli's digital and printable monuments address ethical, philosophical, and historical challenges through an artistic strategy of speculative archaeology and scaling, thereby expanding a historic canon. The artist approaches the original historical artifacts by utilizing fed-in image material, allowing the algorithm to evoke a virtual form of memory from this data pool.

Concluding Remarks

Similar to what Allahyari and Rourke evoke in *3D Additivism*, a wide variety of historical quotations swim in the same stream of sampling, co-existing in their pop-cultural, virtual afterlife – whether in *Mercury* or *Palo Alto*, an antiquizing self-portrait of Robin Rutenberg, or in the virtual reinterpretations and remixes of prehistoric artifacts by Juan Covelli and his artist colleagues. The quasi-timeless, thus ennobled antiquity functions here as a motivic projection surface, from whose (emptied) pool of forms the artists (arbitrarily) draw upon in favor of the entanglement of different worlds. In VR, various material simulations are being experienced. The appropriation of antiquity or an antique-style vocabulary of forms, its proclaimed afterlife and continuity, oscillate between iconoclasm (via fragmentation and uprooting through the transfer to VR) and humorous, playful alienation, material fakery, and conservatism. Playful alienation manifests itself in two ways: On the one hand, the historicity of the respective work mirrors the artists' playful experimentation with the technical possibilities of this (relatively new) medium VR. On the other hand, participatory interaction in a virtual playing field allows for a reception-aesthetic, time-based experience that does not imply the artwork as a pure, static counterpart.

In the form of historical artifacts, history is elastically stretched and drawn into the digital sphere. While Banz & Bowinkel employ a quotation within a quotation, enabling clear identification and attribution, in the works of Covelli and Rutenberg the ancient model can only be stylistically speculated upon;

41 Ibid.

the historical model is alienated and re-created. Covelli claims the digital space as a decolonial space and uses tools such as his computer, VR goggles, and machine learning to contribute to a greater visibility of endangered, inaccessible, or destroyed historical artifacts. He points out the power of the image as such, thereby considering the artistic and speculative forms of sculptural reconstruction in *Speculating the Fragmented Body* as an artistic gesture of empowerment:

> »The problem is what happens with the digitization of these artifacts, they become data, and data is power. Corporations are keeping this data away or are using it to create new sources of profit. These new ways of colonialism mimic older forms of colonialism where you discover, collect, and then profit from it.«[42]

Banz & Bowinkel's choice of subject – Mercury, the god of trade, communication, and thieves – also points to the inherent power structures of big data and the circulation of digital commodities. One of the central stylistic features is the dissolution of fixed contours: fading, crumbling, morphing, and splintering are modes of movement that emphasize the aesthetics of fluidity in »the era of the digital rhizome.« One could ask to what extent these works reflect technological potential itself or whether they refer to an anachronistic concept of sculpture. Antiquity functions as a motivic mediator between the past, canon/tradition, historical continuity, and the present or future. It suggests a reconnection to existing values, but also creates a contrast considering its digitally floating sitelessness. Inscription within an established genealogy is an essential element of Western art history, such as the reception of antiquity in the Baroque period, the use of spolia and the imitation of ancient architectural elements in everyday houses. The previously mentioned artists raise this question about the afterlife of antiquity and its different ways of citation and appropriation. They examine how speculative digital reconstructions reveal both the technological and artistic potential of restoring and reinterpreting collective cultural memories within a speculative-archaeological framework.

42 Ibid.

Bibliography

Banz & Bowinkel in an interview with Dominik Busch (2017), in: Ina Neddermeyer/Claudia Emmert (eds.), Brave New Worlds. Virtual Realities in Contemporary Art, Ex. Cat., Friedrichshafen: Zeppelin Museum, p. 59.

Covelli, Juan in an interview with Schloss-Post (2018): »Ancient Gods and Digital Power«, in: Digital Culture, Theory & Art 0, online: https://schloss-post.com/ancient-gods-digital-power/ (last access: 13.10.2025).

Diprose, Rosalyn (2017): »Speculative research, temporality and politics«, in: Alex Wilkie/Martin Savransky/Marsha Rosengarten (eds.), Speculative Research. The Lure of Possible Futures, London/New York: Routledge, pp. 39–51.

Félibien, André (1699): Des Principes de l'Architecture, de la Sculpture et de la Peinture et des autres Arts qui en dépendent: avec un Dictionnaire de termes propres à chaucun de ces Arts, Paris (Vol. 3), pp. 442.

Focillon, Henri (2018): In Praise of Hands, New York: Parkstone Press International.

Fricke, Beate/Ann-Sophie Lehmann (2024): »Materials Matter«, in: Kunstchronik. Monatsschrift für Kunstwissenschaft 77/7, pp. 461–467. DOI: https://doi.org/10.11588/kc.2024.7.105329

Getsy, David J. (2014): »Acts of Stillness: Statues, Performativity, and Passive Resistance«, in: Criticism 56/1, pp. 1–20.

Gramlich, Naomie (2020): »Feministisches Spekulieren: Einigen Pfaden folgen«, in: Marie-Luise Angerer/Naomie Gramlich (eds.), Feministisches Spekulieren: Genealogien, Narrationen, Zeitlichkeiten, Berlin: Kulturverlag Kadmos, pp. 9–32.

Grau, Oliver (2001): Virtuelle Kunst in Geschichte und Gegenwart: Visuelle Strategien, Berlin: Reimer.

Grötker, Ralf (2001): »Virtuelle Realität«, in: Joachim Ritter et al. (eds.), Historisches Wörterbuch der Philosophie 11, Basel: Schwabe, pp. 1066–1068.

Gustin, Melissa L. (2024): »Do Sleeping Shepherds Dream of 3D-Printed Sheep: John Gibson, Oliver Laric, and Digital Neoclassicism«, in: British Art Studies, London/New Haven: Paul Mellon Centre for Studies in British Art and Yale Center for British Art, online: https://britishartstudies-24.netlify.app/issues/24/digital-neoclassicism/ (last access: 13.10.2025).

Himmelsbach, Sabine (2021): »Unbound Spaces – Immersive Virtualities«, in: Tina Sauerländer (ed.), Resonanz der Realitäten, VR Art Price of DKB in

Cooperation with CAA Berlin, Ex. Cat., Berlin: Haus am Lützowplatz, pp. 20–25.

Kerckhove, Derrick de (1993): »Touch versus Vision: Ästhetik neuer Technologien«, in: Wolfgang Welsch (ed.), Die Aktualität des Ästhetischen, Munich: Wilhelm Fink, pp. 137–168.

Knebel, Sven K. (2001): »Virtualität«, in: Joachim Ritter et al. (eds.), Historisches Wörterbuch der Philosophie 11, Basel: Schwabe, pp. 1062–1066.

Meyer, James (2004): »No more Scale: The Experience of Size in Contemporary Sculpture«, in: Artforum 42/10, online: https://www.artforum.com/features/no-more-scale-the-experience-of-size-in-contemporary-sculpture-168860/ (last access: 13.10.2025).

Michalski, Ernst (1996): Die Bedeutung der ästhetischen Grenze für die Methode der Kunstgeschichte, Berlin: Gebr. Mann Verlag.

Paterson, Mark W. D. (2005): »Digital Touch«, in: Constance Classen (ed.), The Book of Touch, Oxford/New York: Berg, pp. 431–436.

Rutenberg, Robin (2022): Stonesong. A VR Soundscape and Poem, Sound Studies and Sonic Arts Master Exhibition, Berlin University of the Arts, Collegium Hungaricum Berlin, booklet.

Springer, Anton (1886): »Das Nachleben der Antike im Mittelalter,« in: Id. (ed.), Bilder aus der neueren Kunstgeschichte (Band 1), Bonn (2nd, revised and expanded edition).

Ströbele, Ursula (2023): »Sculpting Digital Realities. Notes on Truth to Materials, the Aesthetic Limit, Site-Specificity and 3D-Printing«, in: Mara-Johanna Kölmel/Id. (eds.): The Sculptural in the (Post-) Digital Age, Berlin/Boston: Walter de Gruyter, pp. 83–101.

Ströbele, Ursula (2024): »Elastikakt – Elastizität als künstlerisches (Über-)Lebensprinzip«, in: Julia Wallner (ed.), Der Die Dada. Unordnung der Geschlechter, Ex. Cat., Munich: Hirmer pp. 38–45.

Warburg, Aby (2012): »Der Bilderatlas Mnemosyne«, in: Gesammelte Schriften. Vol. II.1. Ed. by Martin Warnke/Claudia Brink, Berlin: Akademie Verlag.

Weigelt, Shirin (2019): »Tasten: Taktilität als Paradigma des Digitalen«, in: Rainer Mühlhoff/Anja Breljak/Jan Slaby (eds.), Affekt Macht Netz. Auf dem Weg zu einer Sozialtheorie der Digitalen Gesellschaft, Bielefeld: transcript, pp. 107–128. DOI: https://doi.org/10.25969/mediarep/13223

Wilson, Eva (2016): »Hinter den Spiegeln. Virtualität, Rekursion und virtuelle Bilder im 19. Jahrhundert«, in: Heide Barrenechea/Marcel Finke/Moritz Schumm (eds.), Periphere Visionen. Wissen an den Rändern von Fotografie und Film, Paderborn: Wilhelm Fink, pp. 97–112.

Image Credits

fig. 1: © Morehshin Allahyari & Daniel Rourke
fig. 2: © Banz & Bowinkel
fig. 3: © Banz & Bowinkel
fig. 4: © Robin Rutenberg
fig. 5: © Banz & Bowinkel
fig. 6: © Juan Covelli

Curating *Immersion. The Origins, 1949–1969*

Behind the Scenes and Some Afterthoughts

Choghakate Kazarian

I must admit I was surprised to be invited to speak about the exhibition *Immersion. The Origins, 1949–1969* (that I co-curated with Camille Lévêque-Claudet)[1] at a symposium on Virtual Reality. I understand this to be a computer-mediated system that constructs immersive environments to project participants – primarily through visual means – into a simulated world, in contrast to our three-dimensional, tactile reality. It came as a surprise because the exhibition was prompted in response to virtuality and the multiple, online ersatz of real-life experiences implemented during COVID-19 at various levels of our physically constrained lives. The exhibition project was born out of a discussion with Lévêque-Claudet, who was angered by exhibitions such as *Immersive Van Gogh* or *Immersive Klimt* that garnered a huge enthusiasm and that he felt denatured those artists' works and overshadowed the existence of artworks that were immersive by nature. The problem with such spectacles was not their immersivity but the fact that they obtained such a feature by merely projecting distorted and altered images of bidimensional paintings that were never meant to be immersive. Meanwhile, many artists of past generations have made works that are inherently immersive: physical constructions that enclose the viewer, rather than two-dimensional works digitally cast into a wraparound display. For example, the Argentine Italian artist Lucio Fontana realized the pioneering *Ambiente spaziale a luce nera* in 1949, a work that I exhibited in 2014 in the retrospective of the artist that I curated at the Musée d'Art moderne de Paris.[2]

1 *Immersion. The Origins: 1949–1969*, Musée Cantonal des Beaux-Arts, Lausanne (Switzerland), 4.11.2023-3.3.2024.

2 Fontana's pioneering contributions to the genre (that he called *ambienti spaziali*) were later the object of an entire exhibition: *Lucio Fontana Ambienti/Environments* at Hangar

Not only did we feel the need to restore some justice by looking at properly immersive artworks but this act of historical correction also answered a more general need for IRL physical and polysensory experiences that we were deprived of during covid, when everything (experiences and human interactions) was reduced to visual or textual information delivered through screens.[3] The exhibition was thus born in opposition to virtual reality: we wanted to offer an experience of the here and now instead of projecting visitors into somewhere else (the geographically elusive space of internet).

From the outset, our project positioned itself against the logic of substitution that undergirds most virtual reality: the promise to replace here with elsewhere. Our wager was that immersion need not be a digital prosthesis but a condition of the body in space – a compound of friction, weight, temperature, smell, resistance, and social proximity (fig. 1). The exhibition did, however, share certain features common to virtual reality, such as immersivity (the core concept of the exhibition) and interactivity. To some extent, immersive art installations also propose an alternate reality, not through a simulated environment as in virtual reality but by constructing a space within the space of the museum disconnected from its real-world context.

This is how our curatorial proposal came to be associated with virtual reality in the symposium – as a sort of prehistoric, predigital, low-tech forerunner that challenged the contemplative process of viewing traditional art forms. Several works read, in retrospect, like proto-VR interfaces: Gianni Colombo's *Spazio elastico* diagrammed a navigable, wireframe-like coordinate space while USCO's *Fanflashtic* orchestrated multi-sensory cues to tip visitors into a different attentional regime. In this sense, the exhibition functioned as a laboratory for VR's core ambition – how to suspend the ordinary and intensify presence – while insisting that such intensification could remain resolutely IRL, with all the contingency, risk, and reciprocity that physical co-presence entails.

Bicocca, 21.09.2017-25.02.2018, curated by Marina Pugliese, Barbara Ferriani, and Vicente Todolí.

3 IRL (an acronym for In Real Life) has been in use since covid to distinguish reality from the online world.

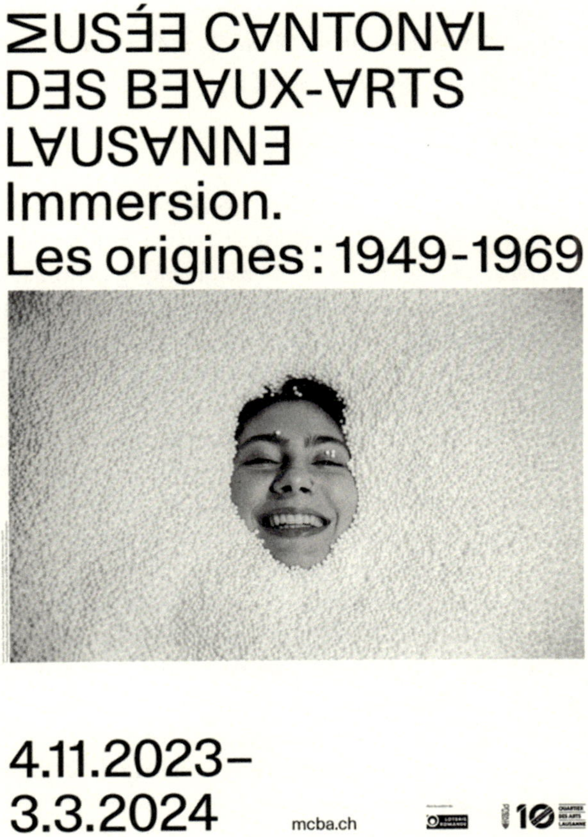

fig. 1: Poster for the exhibition Immersion. Les origines: 1949–1969

In the following essay, I bring up the often invisible processes that underlay the production of the exhibition, revealing the curatorial decisions and logistical challenges that are necessarily absent from the official catalogue, which was completed prior to the exhibition. By foregrounding these behind-the-scenes dynamics, I aim to reveal the contingent nature of exhibition making, highlighting how meaning is constructed not only through the works themselves but also through the labor and choices that organize their presentation and their activation by the public. Furthermore, I offer some afterthoughts on the exhibition's reception by the public in order to explore the dialogic relationship between curatorial intent, institutional framing, and audience engagement.

1. Framing Immersivity

Our initial curatorial task was to articulate a precise understanding of immersivity. While the concept is predominantly associated with computer-generated virtual reality and other forms of entertainment, its examination within the strictly artistic domain remains limited.[4] In art studies, immersivity surfaces in discussions of large-scale installations, yet it is treated as a contingent rather than a systematic or defining feature. Our perspective, that consisted of defining through a specific feature (of current concern), allowed us to look at the emergence of the genre across various movements including Spatialism, Kinetic Art, Zero, Arte Povera, etc. We characterized immersion as the experience of being *inside* rather than merely *in front of* a work. Immersion implies crossing a threshold between two worlds or states, a rupture that suggests a state of full engagement, akin to being submerged in water.[5] This shift relies on a liminal zone marking a transition from the everyday environment of the exhibition space to the world of the artwork. This may be as simple as a curtain that induces darkness, as in Fontana's *Ambiente a luce nera*, or an intermediary corridor, as in Judy Chicago's *Feather Room* or Fabio Mauri's *Luna*.

Defining something is thinking as much about what it excludes as what it includes. Accordingly, we eliminated murals and panoramas, which we considered primarily as painting, as well as architecture, where the act of being within is a given. Projections and light shows were also excluded; during the period we were considering, they predominantly offered frontal, stage-like experiences, reflecting a traditional viewer-object relationship rather than the fully enveloping, 360-degree experience we were investigating. A more ambiguous category consisted of works that we associated with sculpture in the expanded field rather than genuine immersion due to their object-centered nature – for

4 The most important study is Joseph Nechvatal (1999): »Immersive Ideals/Critical Distances. A Study of the Affinity between Artistic Ideologies Based in Virtual Reality and Previous Immersive Idioms«, PhD Thesis, Newport: University of Wales College.

5 For the definitions behind the concept of the exhibition, see Choghakate Kazarian/Camille Lévêque-Claudet (2023): »Immersion. The Origins: 1949–1969«, in: Id. (eds.), Immersion. The Origins: 1949–1969, Ex. Cat., Paris: Hazan, pp. 6–9.

example, Otto Piene's kinetic sculpture *Licht-Ballet*. More decisive was the criterion of autonomy. Therefore, installations that served primarily as backdrops for happenings were ruled out, since their experiential impact depended on external events or interventions that constituted the primary content. While immersive installations inherently involve performativity through visitor interaction, we excluded works from our scope if they functioned mainly as containers for specific performances rather than offering freely accessible, undirected immersive experiences for all visitors. In this way, our exclusions clarified the specific conceptual stakes of immersivity as a totalizing, self-contained, and physically enveloping phenomenon.

As with any definition, ours is not unequivocal, and we remain aware of the porous boundaries of these categories. Much of our curatorial discussions centered on whether a given artwork was sufficiently immersive to merit inclusion in the exhibition. To highlight the specificity of immersive art, we adopted a deliberately restrained definition, which nonetheless proved to encompass a rich and varied range of works. While further connections could undoubtedly have been explored, such as between immersive installations and happenings or between immersive art and design, particularly in Italy – a country notably prolific in producing such works – these fell outside the scope of our exhibition, necessarily limited by the gallery space. Even within these boundaries, *Immersion* offered ample variety in experiences and subcategories, revealing the depth and complexity of immersive practice.

It was immediately clear that, despite the substantial exhibition space (two floors totaling approximately 1,200 square meters), we could only present a limited number of works because of their size. Consequently, we chose to restrict our focus to a specific time frame. In contrast to anachronistic exhibitions such as *Immersive Van Gogh*, we decided to concentrate on the period during which such works emerged in this first overview dedicated to immersive art.

The exhibition's chronological framework was determined both by developments within the art world and by broader sociocultural concerns. We focused on the formative, experimental years of immersive art, a period in which the very notion of space – physical, perceptual, and conceptual – was being actively redefined. Within the art context, the timeline extended from Fontana's 1949 *Ambiente spaziale a luce nera* (fig. 2), which we considered the conceptual and chronological starting point of the exhibition, to the 1969 *Spaces* exhibition at MoMA, which marked the genre's formal institutional recognition and its subsequent proliferation. This time in art history coincided with the »Space

Age« (culminating in 1969 with the Apollo 11 mission on the moon), a period that was deeply immersive in scope and aesthetics, as I demonstrated in my essay for the catalogue.[6] Science, politics, and aesthetics were all preoccupied with space: Cold War politics pursued literal conquest, while artists sought to transcend traditional categories of painting, sculpture, and architecture. By situating immersive art within this dual context, the project underscored its historical specificity and its engagement with the expansive, exploratory ethos of the era.

Although the immersive quality of artworks is not unique to the period under study, it is precisely during this moment that immersive installations emerged as an independent genre, although without being named as such. Terminology varied across linguistic spheres during that period: Latin languages favored »ambiente« (following Fontana's 1949 precedent), while Anglophones preferred »environment«, often associated with Allan Kaprow. This convergence was made explicit at *documenta* 4 (1968), which devoted a section to different kinds of installations – including immersive ones – under the rubric »Ambiente – Environment.«[7] There is, of course, a prehistory of the field: immersive experiences can be traced back to Renaissance trompe-l'œil grottoes, as discussed by Camille Lévêque-Claudet in the exhibition catalogue, or to cave paintings, as analyzed by Joseph Nechvatal in his PhD thesis on virtual reality and immersivity within art.[8] However, we excluded these earlier experiments (as they were part of larger projects), as well as other apparent precedents such as Marcel Duchamp's installation with strings for *First Papers of Surrealism* in 1942, for it functioned as exhibition design intended to contain or frame other artworks rather than exist as an autonomous work. We also excluded Kurt Schwitters's *Merzbau* (1923-1937), a work that occupies a liminal space between sculpture and architecture. In contrast, fully immersive and autonomous works emerged during the period 1949-1969, arising within the cultural context of the »Space Age« and the neo-avant-garde. These works are marked by a conscious departure from traditional artistic categories

---

6 Choghakate Kazarian (2023): »Totalizing Experiences: Immersive Artworks between 1949 and 1969«, in: Ibid., pp. 34–49.

7 For the main concepts used to define such works, see C. Kazarian/C. Lévêque-Claudet (eds.): Immersion.

8 Camille Lévêque-Claudet (2023): »Theaters of Immersion: Immersive Art Exhibitions between 1949 and 1969«, in: Ibid., pp. 14–33; cf. J. Nechvatal: Immersive Ideals/Critical Distances.

and a commitment to pluridisciplinary experimentation and participatory practices.

Considered our historical ambition, we deliberately chose to focus on examples within our chronological framework that materialized rather than on those that remained mere projects.[9] Although unrealized projects were fully situated within the immersive paradigm of the period, including them would have expanded the scope of the study to a scale that was unmanageable within the physical constraints of an exhibition. Moreover, attempting to actualize such projects would have raised significant issues regarding historical accuracy and risked introducing anachronisms.

2. Variety of Immersive Experiences

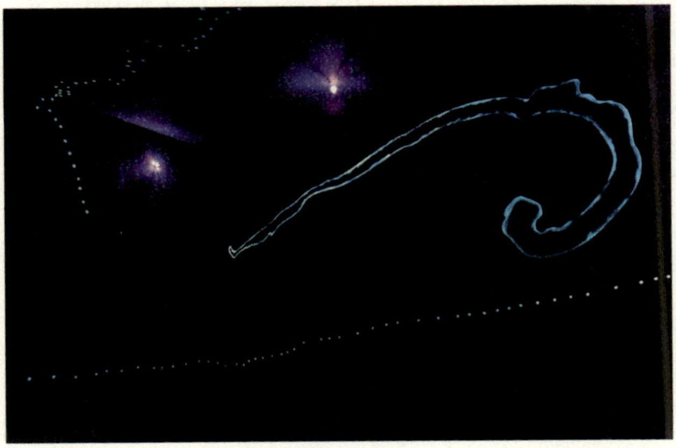

fig. 2: Lucio Fontana: Ambiente spaziale, 1967, installation view: Immersion. The Origins: 1949–1969, Musée cantonal des Beaux-Arts de Lausanne, 4.11.2023-3.3.2024

9 One exception was James Turrell's 1969 *Raemar Pink White*. We discovered a bit late that the project was materialized after 1969, hence its exclusion from the catalogue even though it was presented in the exhibition.

Rather than selecting artworks as mere illustrations of a curatorial argument, we adopted a panoramic approach, showing a representative sampling (a process inherent to any exhibition) of the diverse expressions of the immersive genre within the defined chronological framework. Our selection encompassed works offering a broad spectrum of aesthetics and materials, each proposing a distinct mode of experience.

While firmly anchored in physical media, many of these installations pursue a sense of immateriality. For instance, Lucio Fontana's spatialist work was conceived as drawing in space (fig. 2), while Christian Megert's 1968 *Environment* creates the illusion of infinity through facing mirrors (fig. 3). Whereas most works play with the expansion of space, the exhibition also included an example of claustrophobic immersion in Robert Morris's *Passageway,* a narrowing curved corridor. Some works are overtly spectacular, while others appear deceptively austere and barely perceptible, privileging subtle sensory shifts, such as Laura Grisi's *Vento di s. e. velocità 40 nodi* (a sensation of strong wind created by hidden fans). The unsettling atmosphere of Bruce Nauman's *Sound Breaking Wall,* in which hidden speakers whisper hardly audible sounds of breathing and laughter through the walls, sharply contrasted with the exuberant energy of USCO's *Fanflashtic,* combining strobe lights, balloons, and festive music. To present a broad spectrum of experiences, we limited the selection to one work per artist and avoided including similar pieces. For example, although several artists engaged with the reflexive properties of mirrors, we chose to feature only the Megert work (fig. 3).

We also sought, insofar as possible, to ensure diversity in terms of geographic origin and gender. In constructing a panorama of experimental practices in immersivity, we gave particular attention to foregrounding the contributions of women artists, whose work has often been marginalized or omitted from historical narratives. This objective was supported by the substantial body of scholarship produced in recent decades on women's artistic practices, including important research on figures such as Marinella Pirelli and Laura Grisi. Nonetheless, the number of women represented in the exhibition fell short of our initial intentions. The chronological parameters of the exhibition, situated prior to the emergence of the feminist movements of the 1970s, coincided with a period in which women were afforded fewer opportunities to exhibit, particularly in media demanding substantial spatial and material resources. Immersive installations, by virtue of their scale, were generally conceived for public contexts, a condition that disproportionately favored male artists who benefited from greater opportunities. Furthermore, the process of securing loans for works by women artists proved especially challenging, as such works are currently in high demand. These circumstances contributed to the regrettable absence of pieces by figures such as Yayoi Kusama and Lygia Clark.[10] The catalogue allowed us to include these contributions by women artists to the history of early immersive art.

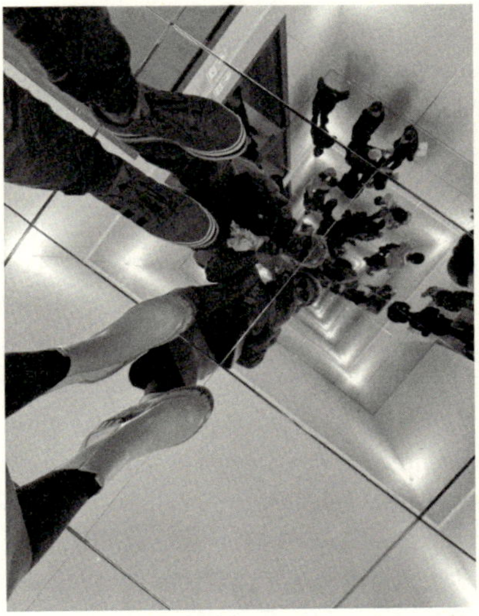

fig. 3: Christian Megert: Environment, 1968, installation view with visitors: Immersion. The Origins: 1949–1969, Musée cantonal des Beaux-Arts de Lausanne, 4.11.2023-3.3.2024

10 While developing the exhibition, we learned that the Haus der Kunst in Munich was preparing a similar project focused on women artists, curated by Andrea Lissoni and Marina Pugliese. Their exhibition spans a much broader period, benefiting notably from the increased presence of women artists in the 1970s. See Andrea Lissoni/Marina Pugliese (eds.) (2024): Inside Other Spaces. Environments by Women Artists: 1956–1976, Ex. Cat., Berlin: Hatje Cantz Verlag.

Another decisive factor in our selection was the requirement that each work be experienced in the manner originally intended. In some cases, the lending institutions we approached wanted to impose conditions that would have altered the fundamental nature of the piece, such as requiring that visitors observe the work from a distance without touching or entering it, as occurred with Marta Minujín's 1963 *Chambre d'amour* and Haus-Rucker-Co's 1968 *Environment Transformer/Fliegenkopf*. We had to abandon borrowing these works, as such restrictions would have transformed the immersive pieces into static sculptures, undermining their purpose.

Finally, practical limitations also shaped the selection. Because they demand substantial space, we were not able to consider Niki de Saint Phalle's *Hon*, Lea Lublin's *Fluvio Subtunal*, or Martina Minujín and Rubén Santantonín's *La Menesunda*. Moreover, the financial costs associated with transporting existing elements, constructing new ones, and (as we later discovered) maintaining the artworks throughout the run of the exhibition were considerable, as the inherently participatory nature of these works and the high attendance exposed them to wear and frequent damage. For these reasons, the inclusion of technology-dependent works was necessarily limited (they were generally in the minority, and VR computer technology was still in its experimental stages). Such pieces impose distinct burdens: specialized (often obsolete) hardware, dedicated technical staff, and contingency stocks of parts whose procurement timelines can exceed the exhibition run. From a stewardship perspective, frequent failures risk compromising the artist's intent, while interim substitutions introduce unacceptable departures from the original concept. Coupled with higher operating costs and an elevated risk of downtime in a high-traffic setting, these constraints led us to privilege installations whose immersive effects could be delivered reliably and sustained by the museum's existing technical capacity.

The final selection for the exhibition comprised fourteen works by artists representing different generations, geographic regions, and artistic orientations: Judy Chicago, Gianni Colombo, Lucio Fontana, Laura Grisi, Fabio Mauri, Christian Megert, Robert Morris, Bruce Nauman, Giuseppe Pinot-Gallizio, Marinella Pirelli, Jesús Rafael Soto, Ferdinand Spindel, USCO, and James Turrell. The accompanying catalogue encompassed a broader range of artists and works.

3. (Re)making of

Beyond these conceptual and curatorial considerations, a number of technical factors had to be addressed, foremost among them the feasibility of reconstructing the works. While some of the issues I have outlined are familiar within curatorial practice, the exhibition posed an entirely novel challenge from a technical perspective, as it involved the re-creation of works. As most of the pieces were originally created for specific exhibitions, their existence was inherently ephemeral, although some were subsequently recreated either by the artist or posthumously, with the authorization of the artist's estate.

It is precisely the idea of creating an effect, with limited concerns with originality, materials, or »the artist's hand,« that enables their re-creation. In several instances, however, reconstruction proved impossible, either because the artist had conceived the work as strictly site-specific, precluding future iterations (as in the case of Michael Asher's installations), or because recreating without the artist would fundamentally compromise the work's integrity. This latter concern applied to Yves Klein's *La Spécialisation de la sensibilité à l'état matière première en sensibilité picturale stabilisée* (also known as *Le Vide*), for which the artist spent days painting the interior of the Iris Clert Gallery entirely white, imbuing the space, in his words, with his own artistic sensibility.[11] While the work has been re-created in recent years (with the authorization of the artist's estate), we felt that without Klein's direct involvement, the result would be, quite literally, void. Accordingly, it was represented in the catalogue but excluded from the exhibition. A further limitation arose in cases where insufficient documentation prevented faithful reconstruction; for example, despite extensive research on a work by Jacqueline Nova and Julia Acuña titled *Luz-Sonido-Movimiento*, presented at the Museo de Arte Moderno de Bogotá, we reached a dead end.

When working with artists' estates or museums, we had to follow strict guidelines to reproduce the work as originally presented. We could occasionally incorporate elements of the original installation (as with Pinot-Gallizio's work) or components from previous reconstructions (as with Fontana's piece). In contrast, when working with living artists such as Chicago, the experience

11 Yves Klein, account of the opening of the exhibition »Époque pneumatique, la sensibilité picturale immatérielle à l'état matière première« at the Galerie Iris Clert, typescript with handwritten annotations, Paris, the Yves Klein Archives.

was different, as she had a flexible approach to her work. While we were reluctant to add further elements due to the historical focus of the exhibition, we accepted certain technical modifications that remained conceptually faithful to the original 1966 work such as the use of more seamless wall screens and feathers that were cruelty-free (fig. 4). Ironically, our re-creations required greater financial and technical resources than the original iterations, which were often surprisingly artisanal and low-tech.

This flexibility with materials was characteristic of all reconstructions carried out by the artists themselves, as our research revealed. Some works, such as those by Mauri and Fontana, were re-created by the artists after their initial presentations. Notably, each subsequent iteration differed substantially from the first. These variations either involved small technical refinements to the original realization or through more substantial changes. For example, Fontana's *Ambiente spaziale a luce nera* (1949) was re-created by the artist in 1967. The later version expanded the room's dimensions and replaced the floating papier-mâché shapes with an organic form cut from cardboard, while the walls were covered by lines of phosphorescent painted dots. Although the artist regarded this re-creation as ideal, it diverged materially from the original, illustrating that, for such works, formal and technical specifics are secondary to achieving a particular effect – in this case, the materialization of a drawing in space. Consequently, we presented the 1967 reconstruction in the exhibition, as the 1949 version was unavailable for loan (fig. 2).

Although we did not anticipate the exhibition's extraordinary success (ultimately attracting a historical attendance of 125,000 visitors in a city with a population of about 139,000), we considered in our selection process the capacity of each installation to accommodate multiple viewers simultaneously, limiting works intended for only one or two participants at a time. The exhibition design allowed sufficient intermediary space between works, enabling visitors to circulate, remove shoes, or queue without obstructing other installations. This arrangement also minimized the potential for »noisy« works to interfere with one another. Nevertheless, some level of cross-contamination was inevitable. Visitors exiting Chicago's *Feather Room* (fig. 4) inadvertently spread feathers throughout the exhibition space (and beyond), while polystyrene beads from Mauri's *Luna* (fig. 1) sometimes found their way into Spindel's *Hole in a Home*'s foamy pink folds or up to Lausanne's train station! As far as we observed, visitors appeared unconcerned by these effects; rather, they displayed their feather-covered clothing as tangible traces of their passage, signifying both an initiation and evidence of having engaged with the work, which they further documented through self-photography.

4. IRL Versus VR: Immersivity in the Age of Social Media

Unfortunately, no sustained studies on visitors to the exhibition were conducted; consequently, the following reflections are grounded in observation. *Immersion* attracted an unprecedented number of teenagers, families, and first-time museumgoers, representing a significant expansion of the institution's audience base. This outcome can be attributed to the exhibition's multisensory appeal – particularly resonant in the aftermath of isolation due to covid – along with the innate human desire for tactile experiences. The longing for collective and interactive encounters and the amplifying effect of social media, where visitor-generated images circulated widely, also played a significant part.

Interaction with the works was largely unregulated, with only minimal guidelines in place for certain installations (e.g., the removal of shoes). While visitor engagement often took a playful form (mostly through self-fashioning), these interactions cannot be equated with gaming, as they lacked formalized rules or a fixed modus operandi. The works also lacked any teleological aim: there were no goals to accomplish, scores to accumulate, or levels to progress through. Instead, the works solicited open-ended exploration and sensorial experience, privileging contingency and self-directed movement over task-oriented achievement – with the partial exception of Morris's *Passageway*: the tilted corridor channels visitors toward a dead end, prescribing a specific itinerary and culminating in a physical – and psychological – terminus. Nevertheless, implicit boundaries emerged, such as the permissibility of touching certain elements (e.g., the feathers in Chicago's *Feather Room*) while refraining from others (e.g., the wires in Colombo's *Spazio elastico*) where contact could result in damage, a risk that did, in fact, materialize. The tactile emphasis of the exhibition also had spillover effects. Guards reported that some visitors – many of whom were experiencing a museum for the first time – attempted to touch artworks in the permanent collection after visiting the exhibition, influenced by the immersive, tactile engagement they had just encountered. This prompted the museum to install signage at the collection entrance reminding visitors not to touch the works.

Interactivity implies a reciprocal exchange, with the potential to alter the work itself. However, none of the artworks in the exhibition extended interactivity to the level of full reciprocal engagement (as in gaming). In other words, they remained fundamentally unaltered by the presence or actions of the audience. Visitors, while able to touch, photograph, or otherwise physically engage with the installations, were not positioned as co-authors of the artworks. Their actions did not meaningfully modify the works' overall compositions. While visitors' actions, such as moving polystyrene pebbles, feathers, colored balloons, or metallic sticks, did affect the spatial arrangement of certain installations, these changes did not fundamentally transform them. When alteration did occur, it was typically unintentional and detrimental, resulting from wear and tear due to the high volume of visitors. Examples include broken screens in Pirelli's *Filmambiente*, damaged wires in Colombo's *Spazio elastico*, feather loss in Chicago's *Feather Room* (fig. 4), and makeup stains on Nauman's *Sound Breaking Wall* left by visitors who pressed their faces against it to hear the subtle sounds it emitted.

Although issues related to high attendance had been anticipated, we had not foreseen the degree of physical wear. This was largely due to the fact that most works were originally conceived for limited audiences and short exhibition durations, ranging from a single night to a few weeks, rather than the sustained four-month run of this exhibition. Consequently, the installations required frequent maintenance to preserve their intended form, and those attending later in the day often encountered works already visibly worn from earlier interactions, underscoring the temporal contingency of the exhibition experience. You cannot step in the same river twice.

While the haptic experience was undoubtedly instrumental to the exhibition's success, particularly in making it accessible to children, the principal factor accounting for its popularity and the diversification of its audience, especially among younger demographics, lay in the reflexive nature of the experience. Although only Megert's mirror installation was reflexive in a literal sense (fig. 3), other works similarly encouraged such engagement through their photographic potential; they served as ideal backdrops for selfies and were subsequently disseminated on social media. Hence, interest accrued for Chicago's *Feather Room* and Mauri's *Luna*, the most popular works on social media.

As noted earlier, the curatorial intention was to reaffirm the importance of physical, multisensory engagement in the wake of the sensory deprivation experienced during covid and to reassert the value of in-person (»in real life,« as they say online) encounters over disembodied, predominantly visual interactions of online experience. The willingness of visitors to queue for up to three hours, despite the extensive availability of images online, suggests that this aim was, in one sense, fulfilled. Yet, this return to the here-and-now was paradoxically underpinned by the logic of social media: the IRL experience functioned simultaneously as an end in itself and as a means of generating content in the virtual world of social media, in which lived experience is instrumentalized for representation. The necessity to be »here« was sustained by a »somewhere else« in the projected world of social media.

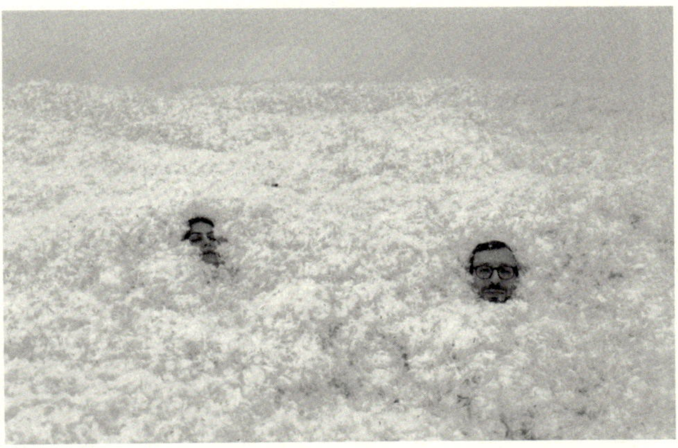

fig. 4: Judy Chicago: Feather Room, 1966, installation view with Choghakate Kazarian and Camille Lévêque-Claudet: Immersion. The origins: 1949–1969, Musée cantonal des Beaux-Arts de Lausanne, 4.11.202-3.3.2024

In translating the embodied, IRL encounter to platforms such as Instagram, the immersive experience of the works was frequently reduced to static images (or short videos). This transformation altered and at times inverted the perception of the works. For example, the numerous Instagram posts of visitors in Chicago's *Feather Room* convey a mood of dreamlike ethereal lightness and immateriality. But IRL, the physical experience was considerably less smooth, as Lévêque-Claudet and I experienced ourselves (fig. 4). Upon entering, we were confronted with the blinding glare of light emitted through translucent walls, intensified by the room's overall stark whiteness; the oppressive heat and pungent odor of feathers quickly followed. Full immersion in the feathers, accompanied by the struggle to avoid suffocation, typically led to our exit. Outside the installation, feathers clung stubbornly to hair and clothing, extending the sensory discomfort well beyond the confines of the room. Such experience is anything but the frictionless »clean room« promised by VR. The exhibition existed in a doubled form: as a site of immediate, embodied encounter and as a visual construct in the economy of digital images, two orders of experience whose affective registers could be diametrically opposed, yet which mutually sustained each other in the cultural logic of mediated, networked spectatorship.

5. Immersivity and Critical Distance

The exhibition was a significant success not only in terms of attendance but also for the intensity of public engagement it generated around a historical display of experimental works by artists, some of whom are unfamiliar even to specialists. Crucial to this success was the exhibition's interactivity. Touch established an immediate phenomenological bond between viewer and artwork, yet the most salient mode of interaction unfolded through self-representation mediated by photography and its dissemination via social media. In this respect, the exhibition reactivated the traditional touristic function of photography (the indexical marker of presence, the declaration »I was there«) while accelerating its circulation and feedback through digital platforms. In this process, aesthetic engagement was displaced into digital self-display.

While this heightened engagement can be interpreted as a democratizing gesture, repositioning the visitor at the center of the aesthetic encounter, it also displaced the works themselves, reducing them at times to backdrops for self-fashioning. While interactivity is celebrated as inclusive and emancipatory, it often risks subordinating the work's historical and aesthetic complexity to the spectacle of the visitor's experience – particularly when calibrated for social media, a platform that circulates a narrow range of codified experiences and, owing to its age-skewed participation, eclipses other forms of engagement that remain undocumented and thus invisible. The exhibition's entertaining value, contingent on the dynamics of audience presence and density, further reinforced this ambivalence. The phenomenological experience of the works shifted markedly depending on whether they were encountered in solitude or within a collective setting. In the absence of other visitors (as I had the privilege to experience), the installations foregrounded an affective register of apeirophobia (anxiety associated with spatial vastness). However, when mediated by the presence of crowds – teenagers documenting the experience through photography, or families with children traversing the spaces in convivial excitement – the same installations were transformed into festive arenas of collective energy inherent to shared experience. In no way I am implying here the superiority of the feeling of solitary anxiety over collective joy (something even more valuable in our post-covid times). Rather, I wish to emphasize that the exhibition's meaning was not fixed within the works themselves but co-constructed through the social dynamics of their reception.

As an experience premised on co-presence rather than headset isolation, the exhibition explicitly opposed the solitary, device-mediated model of virtual reality. Meaning accrued in the shared air of the galleries through the frictions of queuing, the choreography of bodies negotiating thresholds, and the incidental acoustics produced by crowds. Several works even exceeded the museum's bounds by leaving tangible residues on visitors' clothing and skin: feathers from Chicago's *Feather Room* and polystyrene pebbles from Mauri's *Luna* migrated into other rooms and out into the city, functioning as indexical tokens of contact. These traces carried home underscored a material surplus particular to IRL immersion, quite unlike the sealed, »clean-room« virtuality of headset-based VR.

At the same time, the exhibition shared with VR a commitment to world-building: each installation constructed a bounded environment within (and partially against) the museum's default spatial logic. Visitors crossed thresholds, underwent sensory re-calibration, and occupied scenarios that bracketed everyday reference – effects analogous to virtual environments, but achieved with low-tech, predigital means (lights, fans, mirrors, sounds, etc.). The crucial distinction lay in contingency and material resistance: haptic contact, heat, odor, friction, and wear were not bugs to be eliminated but constitutive elements of the experience.

A point of misunderstanding concerned the historicity of the works. Many visitors seemed to apprehend them less as archival reactivations than as opportunities for immediate immersive experience. Several factors contributed to this perception. First, the reconstructed installations, lacking the patina of historical objects, appeared new (and, crucially, were *new*). The invitation to touch, still largely taboo within the museum context, further reinforced that sense. We likewise chose to emphasize immersion as a contemporary experiential mode, thereby permitting the works to communicate

– through affect and perception – the visual culture and broader socio-cultural paradigm inscribed in them. The only contextual element was the wall text for each work that briefly informed readers about the author, title, materials, original setting, and context.[12] The deliberate exclusion of archival ephemera (photographs, catalogues, invitations) in favor of reenactment privileged the affective immediacy of the original experience over its documentation. The original was less resurrected than re-authored under present-day experiential regimes. In a way, this misperception may indicate the effectiveness of reactivating these historical works, even though their original context has inevitably shifted, thereby transforming their meaning.

Another unexpected observation was the absence of vandalism in Morris's *Passageway*, a long, dead-end corridor. The unnerving austerity of this work provoked such frustration when it was first created in 1961 that the walls were defaced with graffiti and other insults, including one by the dancer Yvonne Rainer, who wrote, »FUCK YOU BOB MORRIS.« This incident was referenced in the wall text accompanying the work, a potentially risky curatorial choice that might have encouraged similar acts of transgression. Only a single visitor, however, replicated Rainer's action and words, and in very small writing. It was most likely a playful gesture by an art historian or curator who had read the wall text carefully and enacted a controlled form of engagement, signaling knowledge of the historical context rather than literal defiance.

The exhibition also raised questions about the nature of active versus passive spectatorship. Interactivity is often valorized as active engagement, in contrast to the supposed passivity of visual contemplation. Yet visitor behavior suggested a more complex picture. Despite the infinite photographic possibilities offered by the installations, many participants repeated preexisting images encountered on Instagram, reenacting promotional material such as the photograph of a face emerging from the white pebbles of Mauri's *Luna* that was used for the exhibition poster (fig. 1). The »active« participant is thus enmeshed in circuits of repetition, imitation, and reproduction, complicating the valorization of interactivity as emancipatory. In sum, the exhibition revealed the productive contradictions of immersive museological practice – between historical reenactment and contemporary spectacle, between af-

12 Proportionally, an unusually small number of visitors purchased the exhibition catalogue, which further suggests that the show was primarily perceived as an immediate, experiential event rather than as an educational or historically oriented engagement.

fective immediacy and historical distance, between presence as embodied experience and presence as circulated image.

Bibliography

Kazarian, Choghakate/Camille Lévêque-Claudet (eds.) (2023): Immersion. The Origins: 1949–1969, Ex. Cat., Paris: Hazan.

Kazarian, Choghakate/Camille Lévêque-Claudet (2023): »Immersion. The Origins: 1949–1969«, in: Id., Immersion, pp. 6–9.

Kazarian, Choghakate (2023): »Totalizing Experiences: Immersive Artworks between 1949 and 1969«, in: Id./Lévêque-Claudet, Immersion, pp. 34–49.

Lévêque-Claudet, Camille (2023): »Theaters of Immersion: Immersive Art Exhibitions between 1949 and 1969«, in: Kazarian/Id., Immersion, pp. 14–33.

Lissoni, Andrea/Marina Pugliese (eds.) (2024): Inside Other Spaces. Environments by Women Artists: 1956–1976, Ex. Cat., Berlin: Hatje Cantz Verlag.

Nechvatal, Joseph (1999): Immersive Ideals/Critical Distances. A Study of the Affinity between Artistic Ideologies Based in Virtual Reality and Previous Immersive Idioms, PhD Thesis, Newport: University of Wales College.

Image Credits

fig. 1: © Musée cantonal des Beaux-Arts de Lausanne

fig. 2: © Fondazione Lucio Fontana, Milano, by Siae 2025 © Musée cantonal des Beaux-Arts de Lausanne/Étienne Malapert

fig. 3: photo: Haïkouhie Kazarian

fig. 4: © Musée cantonal des Beaux-Arts de Lausanne/Étienne Malapert

On Dani Ploeger's *Hysteresis*

Jens Schröter

Introduction

This chapter is about a VR art installation called *Hysteresis* (2022) by Dani Ploeger and I want to relate this piece to three different genealogies, namely that of vehicle simulation, that of car pop culture and finally that of the fog machine.[1] I want to start with a helpful statement from Dani Ploeger's website:

> »HYSTERESIS consists of a bumper car that was modified in collaboration with drivers in Zeeland, a rural region in the Netherlands. The car is equipped with a virtual reality headset and accompanied by a video projection. After a virtual race in a Golf Mk4 along a dike road, a tuner talks about his car. Meanwhile, a video of a burnout is projected, repeating endlessly with a Mk4. Burnouts are performed to warm up the tires for a drag race, as a spectacular public intervention and in memory of fellow tuners who have died in accidents.«[2]

Or as Annette Urban, Julia Reich and Manuel van der Veen describe it:

> »In *Hysteresis* (2022), Dani Ploeger consolidates this merger between different media and between different senses: in addition to a VR in a bumper car, he installed incense burners, a video loop showing a Golf Mk4 doing a burnout, and a fog machine that enveloped visitors in smoke. Sitting in the

1 Regarding this work I have to make a disclaimer – unfortunately, I was not able to see it installed, that is why I have to take recourse to still images, videos and descriptions. See the website of Dani Ploeger for helpful materials and videos: https://www.daniploeger.org/hysteresis (last access: 10.07.2025).

2 Dani Ploeger (2022a): »Hysteresis«, in: daniploeger.org. Online: https://www.daniploeger.org/hysteresis-statement (last access: 10.07.2025).

bumper car and with the VR goggles over the eyes, one found oneself, as expected, in the passenger seat of an accelerating car, while sound, smell and wet fog bridged the distance between the worlds. The virtual ride ends with a violent crash into a tree, which Ploeger uses to provoke the abrupt end of the passthrough. But despite the physical shock, visitors are not simply catapulted out of the VR; instead, they look back at the exhibition ensemble, which at first seemed intrusive, as if at a memorial ceremony for the virtual accident.«[3]

In the following I want to analyze this installation, especially regarding what it might say and show in relation to Virtual Reality. At first, it has to be remarked that the installation evokes at least three different genealogies – except from the history of installation art as such (which I will mention later, albeit shortly):

1. The genealogy of vehicle simulation
2. The genealogy of, what I call in lack of a better term, car pop culture
3. The genealogy of the fog machine.

In 4) I will describe the several structural oppositions that are central to Hysteresis. In 5) I will draw a short conclusion.

1. The genealogy of vehicle simulation

The use of a head-mounted display (HMD) in which a virtual ride is displayed while sitting in a real vehicle seems to quote the genealogy of vehicle simulation which has been a very important technology for many years. Many authors start the genealogy of virtual reality with flight simulation.[4] In the highly technologized world of modernity a ›simulation space‹, a simulation infrastructure, must partially and approximately duplicate this world – in order for this world to function at all, because we are surrounded by high-risk technologies.

3 Julia Reich/Annette Urban/Manuel van der Veen (2023): »passthrough. Von Portalen, Durchblicken und Übergängen zwischen den (virtuellen) Welten«, in: kunstforum.de (07.2023). Online: https://www.kunstforum.de/artikel/passthrough/ (last access: 10.07.2025), n.p., my translation.

4 See for example the early book by Benjamin Woolley (1992): Virtual Worlds, Oxford: Blackwell Publishers, Ch. 2.

Airplanes, nuclear power plants, high speed trains and many more can only be used by trained subjects – otherwise catastrophes might occur.[5]

As Virilio emphasized, every new technology is inevitably accompanied by a new type of accident – like the car accident shown in *Hysteresis*. It is therefore equally to be expected that additional techniques will be invented to prevent accidents or at least make them less likely – simulators are such techniques, but Virilio does not mention them.[6]

The history of flight simulation is mostly dated back to the »Link Trainer«, invented around 1931.[7] Jeon remarks on simulative subjectivity:

> »[The] Link Trainer connected flight as action and flight as representation – flying and ›flying‹ – providing a space for the two forms to shape each other. [...] Again, a new kind of pilot was implicated in the practice of instrument flying: those who could ›believe in their instruments rather than themselves.‹ It meant a transition from ›natural pilots‹ with feel and intuition to ›mechanical pilots,‹ [...] For almost five years after its first invention, however, Link's own flying school was the sole user of the trainer for flying instruction. A majority of its early sales went to amusement parks, and there were only a few purchases from aviation-related organizations.«[8]

Jeon mentions an important point, that already played a role in Link's patent. Link introduced the trainer, besides many other potential uses, as »entertainment apparatus for profit«.[9] And obviously a lot of the entertainment media today, think of certain types of computer games, are based on simulations.

---

5 Jens Schröter (2025): »Fictional Infrastructures«, in: Markus Gabriel/Marion Gymnich/Birgit Ulrike Münch (eds.), Wirklichkeit/Fiktion. Spannungsfelder eines Verhältnisses, Tübingen: Mohr-Siebeck, pp. 261–275.

6 Paul Virilio (2005): L'accident originel, Paris: Éditions Galilée.

7 See B. Woolley: Virtual Worlds, Ch. 2.

8 Chihyung Jeon (2015): »The Virtual Flier: The Link Trainer, Flight Simulation, and Pilot Identity«, in: Technology and Culture 56/1, pp. 28–53, here: p. 30 and 33.

9 E. A. Link Jr. (1931): »Combination Training Device for Student Aviators and Entertainment Apparatus«, US Patent 1,825,462.

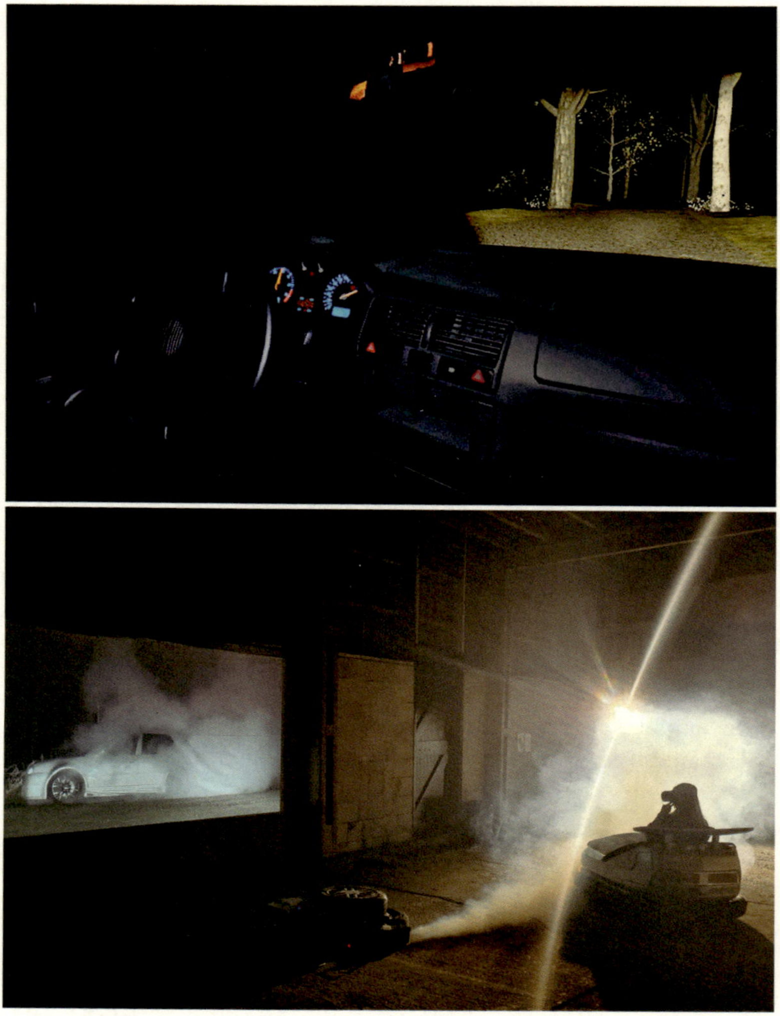

fig. 1: Dani Ploeger: Hysteresis, 2022, modified bumper car, VR headset, video projection, smoke machine, car tires, burnt rubber (above: VR view; below: installation view)

Virtual Reality (VR), in a sense, emerges from such simulators, as its name already suggests. It is connected to reality, but it is virtual at the same time.

A simulated plane or car must behave like the real one, otherwise you cannot be trained for the real use. However, if you have an accident – as in the video in the HMD in Ploeger's installation – you should not die, otherwise simulators are useless. What we call VR today is mainly the fictionalization of simulated space, that is virtuality and fictionality are combined.[10] The reality effect is used, but combined with fictional content. That is basically still the dispositive of many VR artworks.

Although the HMD in Ploeger's work alludes to VR, it is of course not a simulator in any strict sense. The viewer in the bumper car does not steer the virtual vehicle. The person just sits there and watches the immersive video with the HMD. A structural opposition appears here between the real viewer and the implied viewer (because the implied viewer steers the virtual car). This is only the first of a series of structural oppositions that are constitutive for the whole installation, which is a kind of heterogeneous ensemble. Now, there are elements in the installation that obviously do not come from the genealogy of simulators, but clearly more from the genealogy of »amusement devices«.

2. The genealogy of car pop culture

Bumper cars are a well-known and, as far as I can tell, still existing amusement apparatuses. In a way they share some properties with vehicle simulation, although they are obviously not intended as »control environments«[11], as simulators are. They are fun environments. You drive a vehicle around, even bump into other vehicles – which would be an accident in the case of real cars. Bumper cars are not used to train anyone to not bump into other vehicles. Bumping is allowed, but of course that does not mean that no accidents can happen with bumper cars on a secondary level, actually they do.

10 See Jens Schröter (2004): Das Netz und die Virtuelle Realität. Zur Selbstprogrammierung der Gesellschaft durch die universelle Maschine, Bielefeld: transcript, pp. 211–215.

11 S. R. Ellis (1991): »Nature and Origins of Virtual Environments. A Bibliographical Essay«, in: Computing Systems in Engineering 2/4, pp. 321–347, here: p. 327. Ellis means with his notion »control environment« that simulator technology is normally used to train people to use potentially dangerous technologies (like airplanes, nuclear power plants, high speed trains, but nowadays even in some driving schools simulators are used). Simulators are used to control the future usage of such technologies, to discipline subjects into being appropriate users.

Bumper cars may be viewed as part of a popular culture of the car, an important part of second half 20th century pop culture.[12] *A Streetcar named Desire* – such a movie title would not have been possible without that kind of popular culture. Of course, also McLuhan, in the car-crazed US, had to write some pages on the car in *Understanding Media* from 1964, with remarkable sentences like: »The simple and obvious fact about the car is that, more than any horse, it is an extension of man that turns the rider into a superman: It is a hot, explosive medium of social communication.«[13] At least it is true that in certain fan and tuning subcultures cars are an ›explosive medium of social communication‹ (and perhaps also when car drivers direct their aggressions against each other). In these cultures every property of a car is discussed and even a ›burnout‹, the smoking tires we can see in the video for *Hysteresis*, becomes a communicative tool, e.g. being a reminiscence to fellow tuners that have died in accidents. The use of the ›burnout‹ is also interesting because Ploeger wrote a book on »everyday technologies in extreme circumstances«:

> »I have been interested in the ways in which everyday technologies are used and appropriated under circumstances that their designers and manufacturers did not envisage and in ways that do not correspond with their usual representation in advertising and other media. [...].«[14]

Although, unfortunately, Ploeger does not discuss car cultures, simulators, bumper cars or fog machines (I will come to that later) in his book, we can cling to his remark that one of his central interests is:

> »[e]xamining the ways in which technologies that were designed, produced and marketed for use in everyday consumer culture start to (mal)function, gain new meanings and are appropriated in these liminal spaces can give us hints at what technology beyond global consumerism could look like.«[15]

12 See David Thomas/Len Holden/Tim Clayton (eds.) (1998): The Motor Car and Popular Culture in the 20th Century, Aldershot a. o.: Ashgate.

13 Marshall McLuhan (1994): Understanding Media. The Extensions of Man, Cambridge, MA/London: MIT Press, p. 221.

14 Dani Ploeger (2021): Deserted Devices and Wasted Fences. Everyday Technologies in Extreme Circumstances, Axminster: Triarchy Press, p. 1.

15 Ibid., p. 2.

That is: the way in which technologies were designed and how they are appropriated is a relevant perspective to understand his work – although admittedly simulators and bumper cars are not strictly everyday technologies, and an art installation is perhaps not the kind of extreme circumstance Ploeger envisions. Anyway: In the context of Ploeger's cooperation with the Netherlandish tuning community cars are brought into extreme circumstances. And his cooperation includes that community of practice, which is at the same time a fan community, into the installation. Moreover, the tuning of cars is an appropriation and modification of car technology, sometimes even beyond the limits of what is allowed by the law. A burnout is an extreme circumstance for a tire. The tuning community has without doubt a certain transgressive moment, which reminds one immediately of J. G. Ballard's novel *Crash*. This remarkable aestheticization of the car is often linked, because of its topic, people that become sexually aroused by having accidents, to McLuhan's thesis that media are extensions of men or even more, as he writes in his chapter on cars in *Understanding Media*: »[M]en have always been the sex organs of the technological world. The car is no more and no less a sex object than the wheel or the hammer.«[16] The central figure of *Crash*, Vaughan, is a kind of outlandish media studies professor and in one scene the characters view films of »slow-motion films of test collisions.«[17] The narrator speaks also of »simulated collisions we had seen at the Road Research Laboratory.«[18] In *Crash* ›simulated‹ is a quite frequent word, connected to the car-machine and the body-machine at the same time.[19] The exhaustion of the tire in Ploeger's *Hysteresis* is a metaphor for a Bataillian transgressive excess that undermines the purpose of simulation to produce controlled environments and the purpose of bumper cars to allow safe bumping. There is a tension between the symbolic taming of the real and the transgressive reconnection to real danger – the smoke of the burnout tire is a concise sign for this, since smoke has always been seen as an indexical sign for fire and fire is the danger for the stabilized normality par excellence, an all-consuming force that

16 M. McLuhan: Understanding Media, p. 220.

17 J. G. Ballard (1973): Crash, London: Jonathan Cape, p. 8.

18 Ibid., p. 147.

19 Ibid., p. 61, 73 (in relation to a haircut), 85 (in relation to wounds), 97 (»technology of accident simulation«), 98 (»simulated crash«), 133 (skin colours of whores), 140 (juvenile anuses), 147. Sometimes simulation is related to car technology, sometimes to the body. This shows how the machine and the body are part often the same discursive chain for Ballard.

disrupts every order. The represented smoke of the tire obviously connects with another element of the installation: the fog machine.

3. The genealogy of the fog machine

Another part of Ploeger's installation *Hysteresis* is a fog machine.

> »A fog machine, fog generator, or smoke machine is a device that emits a dense vapor that appears similar to fog or smoke. This artificial fog is most commonly used in professional entertainment applications, but smaller, more affordable fog machines are becoming common for personal use. Fog machines can also be found in use in a variety of industrial, training, and some military applications.«[20]

The military uses might be: you can use those machines to hide military structures on the ground by producing synthetic fog, or you can train soldiers on how to orient in difficult situations in which visibility is constrained. But most of us know fog machines from concerts of rock and pop bands and perhaps also from electronic dancefloor clubs etc. Kittler writes: »A fata morgana machine that can now be had around the globe. Without war, simply by paying an admission fee. For mechanization has also taken command over so-called times of leisure and peace.«[21] He then mentions the »discotheque«: »Deaf, mute, and blind, bodies are brought up to the reaction speed of World War n+I, as if housed in a gigantic simulation chamber. […] Every culture has its zones of preparation that fuse lust and power, optically, acoustically, and so on. Our discos are preparing our youths for a retaliatory strike.«[22] Again, the notion of simulation is used – simulators are of course also ›zones of preparation‹. Bumper cars are a perfect example for mechanization taking command over leisure and pleasure. And surely, in the discotheque, the combination of deafening loud music, flickering lights and a fog machine produces a war like atmosphere, in which the senses are overwhelmed by explosions. So, using a

20 Wikipedia (2025): »Fog Machine«, in: wikipedia.org. Online: https://en.wikipedia.org/wiki/Fog_machine (last access: 10.07.2025).

21 Friedrich Kittler (1999): Gramophone, Film, Typewriter, Stanford: Stanford University Press, p. 140.

22 Ibid.

fog machine evokes settings with potentially disturbing effects on the sense organs – another case of »extreme circumstances«.

The fog machine produces a simulation of the smoke of the burnout tire and itself quotes the genealogy of zones of preparation, in which visibility is reduced. By introducing this hindrance to visibility into the installation, Ploeger also thematizes the fundamental problem of all visual art – the relation of the visible and the invisible. It is worth remembering that the cloud, due to its fuzziness, its permanently changing shapes and its role in blocking to the view onto the sun, the moon and the stars – think of Damisch's impressing study *A Theory of /Cloud/* – also marks a special problem in the history of representation.[23]

4. The structural oppositions in *Hysteresis*

After discussing some of the elements that are used in *Hysteresis*, I want to delve deeper into the analysis of the installation. The work is organized around several structural oppositions:

- the opposition between the public visible projected video image and the first-person image in the HMD
- the opposition between the filmed fog of the burnout and the real fog of the fog machine, implying what Urban, Reich and van der Veen rightly pointed out when emphasizing »sound, smell and wet fog bridged the distance between the worlds«[24]
- the opposition between the filmed car and the real bumper car – and implying the opposition between car (›serious technology‹) and bumper car (›amusement device‹). A difference that is repeated in the VR setting between ›serious‹ simulation and (fictionalized) ›simulation for fun‹
- the opposition between the filmed car and the virtual car of the HMD animated video sequence
- the opposition between the space of the video, the simulated space and the real space

23 Hubert Damisch (2002): A Theory of /Cloud/. Toward a History of Painting, Stanford: Stanford University Press.

24 J. Reich/A. Urban/M. van der Veen: passthrough, n.p., my translation.

- the opposition between the bright light illuminating the installation and the filmed light in the video and the implied light in the VR sequence
- the opposition between visibility and invisibility

Why do I stress this complex ensemble of oppositions so much? Because I think that the art form of installation, under which I subsume Ploeger's *Hysteresis*, has a special affinity of exposing several medial oppositions. Installations spatially combine different media and materials, which allows for different spectatorial positions at the same time. One example: The person in the bumper car wearing the HMD and the persons watching the installation from the outside are perhaps asking themselves, if the video they see is the same video that can be seen in the HMD – what is often the case in VR installations.

I would suspect that the whole concept of such installations is exactly to expose the viewers to such oppositions. Bishop speaks of »the multiperspectivalism implicit in installation art.«[25] Otherwise, the co-presence of heterogeneous elements would make no sense. But what might be the specific operations in this case? Since it is a VR installation we can assume that one central concept is to contrast and compare real and virtual space. However, that is in a way true for all such installations that combine a VR presentation, displayed through a HMD, with other elements – and to recall: we can find at least three central themes in *Hysteresis*:

- Simulators as training- and disciplinary-technology, but also as entertainment
- Popular car culture
- The fog machine and its cultural repercussions

The general contrast and comparison of virtual and real is realized here in a specific way: Although the use of HMDs and of virtual space always connotes its genealogy in simulation and simulators, these connotations are combined here with bumper cars, popular car culture and the fog machine. Thereby entertainment and the use of technology against the grain, so to speak, is foregrounded. At the same time, e.g. the contrast between the filmed fog and the real fog underlines the distance between real space and the diegetic space of film as does the HMD in the bumper car underline the distance between real space and the virtual space of simulation. Also, the filmic and the computer-generated space

25 Claire Bishop (2005): Installation Art, London: Tate Publishing, p. 54.

are put into contrast. None of this really fits together: The fog in the video is also chemically different from the fog of the fog machine. The bumper car cannot be steered, as is possible when used as an amusement vehicle; the virtual space witnessed in the HMD is not an interactive space as it normally is. Moreover, the fog from the fog machine temporarily clouds the scenery in the installation and problematizes the role of visibility as such. The fog of the fog machine therefore also evokes, when related to the hyper-perspectival HMD again, the extreme opposition between order and its other. That these different oppositions cannot be smoothly reconciled is not a deficit of *Hysteresis*, but the precise way in which the installation holds open the meanings of the real and the virtual and puts them in constant exchange.[26] We have several – multiperspectival – different subject positions accordingly: We have the driver in the video who burns out the tire, the POV of the camera, we have the person in the bumper car, the POV of the virtual image in the HMD, and finally the observer of the whole installation. In that sense the installation multiplies different points of view. This may be read as a meditation of the complexities of simulative subjectivity.

Conclusion

Finally, the questions of all questions: Why is it called *Hysteresis*? In a text on Ploeger's work we can read:

> »Hysteresis – ›The energy that is lost and not returned when tire materials are stressed in either direction. The lost energy is converted by molecular interaction into heat, and since rubber has poor heat conductivity, the internal temperatures of a tire can increase rapidly‹ (The Automobile Lexicon)«[27].

In a technical paper on »Vehicle dynamics with brake hysteresis« we can read: »Braking is a key action to ensure the safety and stability of vehicle motion; a

26 Cf. Gilles Deleuze (1995): Difference and Repetition, New York: Columbia University Press, p. 100: »We see both that the virtuals are deducted from the series of reals and that they are incorporated in the series of reals.«

27 Dani Ploeger (2022b): »Hysteresis, Dani Ploeger«, in: art-claims-impulse.com. Online: https://www.art-claims-impulse.com/de/hysteresis-dani-ploeger-de (last access: 10.10.2025), my translation.

reaction delay due to hysteresis can be very critical.«[28] Again: the accident and the possibilities to react are centrally relevant. We can interpret hysteresis as a kind of deferred resistance of materials – in that sense it is similar to the notion of feedback, to the answer a system gives to a subject (or to another system), depending on earlier interactions. In that sense, with the title »Hysteresis« the installation points to the resistance of non-human actors that can even be dangerous and at the same time to its own interactive elements. In emphasizing dangerous materiality, the limits of virtual simulation are highlighted. Actually, after a crash in a virtual car race, like the one that can be seen in the HMD in *Hysteresis*, one can start all over again. That is the whole point of simulation – to be reversible. The same is with the fog: The fog from the machine dissipates irreversibly; the fog in the projected image can be repeated again and again. The installation is not just a memento mori for a potentially crashed driver transferred to the gallery space. And why should this be interesting? Rather, I think, it can be read as a complex meditation on the place of virtual space, of simulation, of training, between serious uses and entertainment, between the virtual and the real.

Bibliography

Ballard, J. G. (1973): Crash, London: Jonathan Cape.

Bishop, Claire (2005): Installation Art, London: Tate Publishing.

Damisch, Hubert (2002): A Theory of /Cloud/. Toward a History of Painting, Stanford: Stanford University Press.

Deleuze, Gilles (1995): Difference and Repetition, New York: Columbia University Press.

Ellis, S. R. (1991): »Nature and Origins of Virtual Environments. A Bibliographical Essay«, in: Computing Systems in Engineering 2/4, pp. 321–347.

Jeon, Chihyung (2015): »The Virtual Flier: The Link Trainer, Flight Simulation, and Pilot Identity«, in: Technology and Culture 56/1, pp. 28–53.

Kittler, Friedrich (1999): Gramophone, Film, Typewriter, Stanford: Stanford University Press.

28 Barys Shyrokau/Danwei Wang/Klaus Augsburg/Valentin Ivanov (2012): »Vehicle dynamics with brake hysteresis«, in: Proceedings of the Institution of Mechanical Engineers, Part D: Journal of Automobile Engineering 227/2, pp. 139–150, here: p. 139.

Link, E. A. Jr. (1931): »Combination Training Device for Student Aviators and Entertainment Apparatus«, US Patent 1,825,462.

McLuhan, Marshall (1994): Understanding Media. The Extensions of Man, Cambridge, MA/London: MIT Press.

Ploeger, Dani (2021): Deserted Devices and Wasted Fences. Everyday Technologies in Extreme Circumstances, Axminster: Triarchy Press.

Ploeger, Dani (2022a): »Hysteresis«, in: daniploeger.org. Online: https://www.daniploeger.org/hysteresis-statement (last access: 10.07.2025).

Ploeger, Dani (2022b): »Hysteresis, Dani Ploeger«, in: art-claims-impulse.com (14.09.2022). Online: https://www.art-claims-impulse.com/de/hysteresis-dani-ploeger-de (last access: 04.08.2025).

Reich, Julia/Annette Urban/Manuel van der Veen (2023): »passthrough. Von Portalen, Durchblicken und Übergängen zwischen den (virtuellen) Welten«, in: kunstforum.de (07.2023). Online: https://www.kunstforum.de/artikel/passthrough/ (last access: 10.07.2025).

Schröter, Jens (2004): Das Netz und die Virtuelle Realität. Zur Selbstprogrammierung der Gesellschaft durch die universelle Maschine, Bielefeld: transcript.

Schröter, Jens (2025): »Fictional Infrastructures«, in: Markus Gabriel/Marion Gymnich/Birgit Ulrike Münch (eds.), Wirklichkeit/Fiktion. Spannungsfelder eines Verhältnisses, Tübingen: Mohr-Siebeck, pp. 261–275.

Shyrokau, Barys/Danwei Wang/Klaus Augsburg/Valentin Ivanov (2012): »Vehicle dynamics with brake hysteresis«, in: Proceedings of the Institution of Mechanical Engineers, Part D: Journal of Automobile Engineering 227/2, pp. 139–150.

Thomas, David/Len Holden/Tim Clayton (eds.) (1998): The Motor Car and Popular Culture in the 20th Century, Aldershot a. o.: Ashgate.

Virilio, Paul (2005): L'accident originel, Paris: Éditions Galilée.

Wikipedia (2025): »Fog Machine«, in: Wikipedia.org (22.07.2025). Online: https://en.wikipedia.org/wiki/Fog_machine (last access: 10.07.2025).

Woolley, Benjamin (1992): Virtual Worlds, Oxford: Blackwell Publishers.

Image Credits

fig. 1: Courtesy the artist

The Area of the Virtual

About the Collision of Spatial Perception and Spatial Projection in VR

Manuel van der Veen

This chapter begins with a brief overview of artworks that employ head-mounted displays. The purpose of this overview is to illustrate how the use of VR in the field of art has developed in different directions in recent years. These developments not only build on what has already been done, but also undermine the very foundations of VR, particularly its spectacular immersion of the viewer into an image.[1] Today, there are even cases with VR in which users cannot enter the virtual space at all. Instead, they can only view it from the outside via a second screen. For example, Hayden Fowler's *Together Again* (2017) which takes place in a cage (fig. 1a).[2] By referring to Joseph Beuys' performance *I like America and America likes me* (1974) with a coyote, Fowler got himself locked up with his dingo. The artist wears a headset and different tracking devices are attached to the dingo. Only a small second screen in the corner of the cage provides insights into a virtual desert and thus into what the artist sees through the glasses. While the cage remains invisible to Fowler, the dingo cannot perceive the virtual world. Although the artist and the dingo can still sense each other through touch and smell, and, thanks to the tracking devices, the dingo is represented as a digital double in the desert, both are still trapped: the dingo in the cage and the artist in the hermeticism of VR. Only the viewers are able to watch the entanglement of the two enclosures – beyond immersion. They can look at the screen to see into VR and through the bars of

1 Cf. David J. Chalmers (2022): Reality+. Virtual Worlds and the Problems of Philosophy, New York: W. W. Norton & Company, especially the Ch. »Do virtual reality headsets create reality?«.

2 I would like to thank Julia Reich for bringing this work to my attention.

the cage. In addition to this special case of inaccessible VR, there are site-specific VRs today, in which a digital space is accurately adapted to the exhibition space. Furthermore, there are virtual worlds that still enclose viewers, while at the same time they are perforated with small windows through which the actual surroundings can be seen. There are VRs which include actual objects that can be touched and used and there are works of art that couple VR with augmented reality (AR) and machine learning.

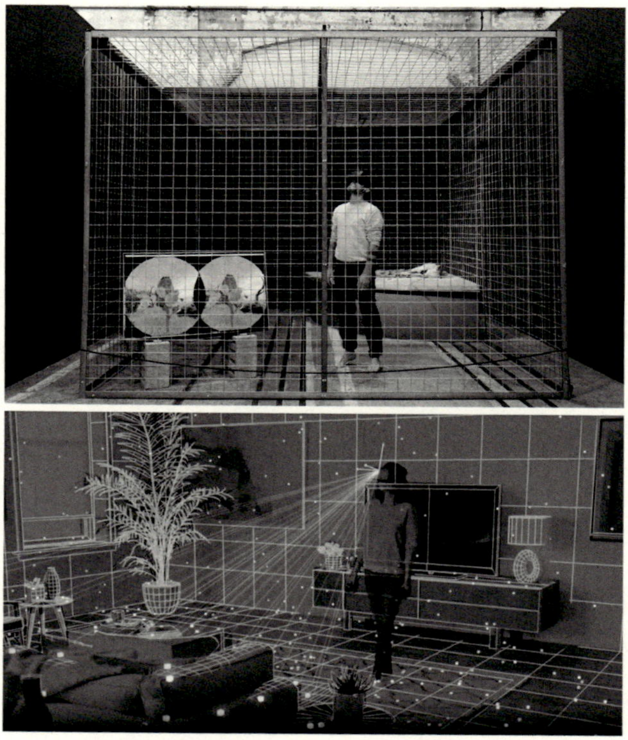

fig. 1a&b: above: Hayden Fowler: Together Again, 2017, installation, performance, HTC Vive, installation view: Sydney Contemporary, Carriageworks; below: Screenshot from the video: Powered by AI: Oculus Insight

These examples are variations of VR that combine different characteristics of extended realities (XR) and experiment with transitions between varying procedures. They indicate a tendency that is even more evident because it has a parallel in technical development. Over the last years, artists have been confronted with headsets that can be used for all XRs due to their integrated *passthrough mode* and advanced spatial sensors.[3] Current glasses make it possible to realise virtual, augmented, and mixed realities with a single device. Therefore, one could ask: what does it actually mean to say VR? What significance does such a concept still have? And does it even make sense to talk about VR when all the examples mentioned above are subsumed under it?

In an updated version of Paul Milgram and Fumio Kishino's virtuality-reality continuum, created by Richard Skarbez, Mary C. Whitton and Missie Smith, VR is completely excluded, as there is no experience that is purely ›virtual‹.[4] However, Hayden Fowler's example cannot be positioned in any of the classification systems, nor do they allow for asymmetries. In addition, it must be said, that VRs that cannot be entered, VRs that are site-specific, or VRs that integrate windows into the exhibition space, all lead to completely different art-historical questions. Paulien Dresscher and Nana Verhoeff even state that the X in XR refers to the great unknown, »as to indicate that the emerging entities cannot be predicted or are not (to be) predetermined.«[5] In order to avoid leaving the status of XR completely uncertain, I would like to introduce my concept of *areal space* in this chapter, which unfolds a fundamental condition for these technologies while maintaining the necessary openness of the virtual. This *areal space* is a concept that addresses the aforementioned ambiguity and allows for a more theoretical and art-historical classification of the various procedures, whereby it also explicitly refers to technical characteristics. In German, ›area‹ is called *Areal* which connects a spatial area with

3 Cf. Julia Reich/Annette Urban/Manuel van der Veen (2023): »passthrough. Von Portalen, Durchblicken und Übergängen zwischen den (virtuellen) Welten«, in: Kunstforum International 290, pp. 86–95, and also see Chris Salter's discussion on the HMD in this volume.

4 Cf. Richard Skarbez/Mary C. Whitton/Missie Smith (2021): »Revisiting Milgram and Kishino's Reality-Virtuality Continuum«, in: Frontiers in Virtual Reality 2, n.p. and see also Paul Milgram/Fumio Kishino (1994): »A Taxonomy of Mixed Reality Visual Displays«, in: IEICE Transactions on Information and Systems E77-D/12, pp. 1321–1329.

5 Paulien Dresscher/Nanna Verhoeff (2020): »XR. Crossing and Interfering Artistic Media Spaces«, in: Larissa Hjorth/Adriana de Souza e Silva/Klare Lanson (eds.), The Routledge Companion to Mobile Media Art, New York: Routledge, pp. 482–92, here: p. 483.

a-reality, i.e. with the unreal. As a matter of fact, the term a-real is not the negation of the Latin word *realitas*. To paraphrase the French philosopher Jean-Luc Nancy, one could nevertheless call this a lucky circumstance that is being exploited here:

> »Areality is an antique word, signifying the nature or specificity of an aire (›area‹). By chance, this word also serves to suggest a lack of reality, or rather a slight, faint, suspended reality: the reality of a swerve localizing the body, or a displacement within the body. In effect, a faint reality of ground, substance, matter, or subject. But this faint reality makes the whole areal real, where the so-called archi-tectonics of bodies is played out and articulated.«[6]

While Milgram and Kishino distinguished between the ontological categories of ›real‹ and ›virtual‹, the updated version is framed between the reception-aesthetic category of ›immersion‹ and the production-aesthetic category of ›extend of world knowledge‹[7]. The concept of *areal space*, in contrast, is based on a technical specificity of current technologies of virtuality in order to proceed with it theoretically and aesthetically.

This specific feature is that all procedures – whether virtual, augmented, or mixed reality – are both space-perceiving and space-projecting.[8] Following this feature means that none of the XR completely excludes the outside space and that none shows the outside space as it is. This is because all XR perceive the surrounding space in a certain way and project a specific spatiality of their own. In VR, for example, the space is constantly perceived via spatial recognition (fig. 1b). Indeed, it is precisely in this process that a 3D model of the actual surroundings is generated, but not for the purpose of archiving it or transporting it to another location. Instead, this very 3D model is invisibly projected onto the environment in order to mediate between human and machine perception. Depending on the situation, the boundaries of the spatial model can either coincide with

6 Jean-Luc Nancy (2008 [1992]): Corpus, New York: Fordham University Press, p. 43.

7 In R. Skarbez/M. C. Whitton/M. Smith: Revisiting Milgram and Kishino's Reality-Virtuality Continuum, the authors mix the modality of experience ›immersion‹ with the aspect of ›extend of world knowledge‹, which refers to what the machine knows about the world.

8 I proposed this feature in my lecture *Placing Things Beside Themselves. ›to see how they hold up‹ (Georges Braque)* for AR and have continued to develop it since then. The talk was given as part of the international conference *THINGS BESIDE THEMSELVES* at Bauhaus University Weimar (Research Group Media and Mimesis) in February 2020.

those of the actual environment, or mark a smaller area within it. For example, with Meta Quest's Guardian, a safety feature, users mark an area around the available space in the room, which then becomes a kind of cage, as in Hayden Fowler's case.[9] The spatial reference to the immediate surroundings is active in all variants of XR. This also applies if the actual surroundings are subsequently obscured by a second projection; in the case of VR, this is the visible enveloping projection of a virtual world. A technical specificity of XRs is their space-perceiving and space-projecting property, which takes place even before virtual objects or virtual worlds appear. Hence, a major part of their aesthetics results from the relationship, dependence and collision between the perception and projection of space. The concept of *areal space* is based on this technical specificity without using any ontological attributions such as being virtual. The concept is proposed here because it allows for comparisons with other art historical concepts, enables individual XRs to be positioned in relation to one another, and provides a basis for theoretical connections.

In the first section, the concept of *areal space* will be developed in general terms, in order to elaborate on a more specific VR area in the second section, which will be discussed with regard to individual artworks that intertwine the virtual area with an exhibition's architecture. The two works discussed, one by Karin Sander and one by Theodoulos Polyviou and Eleni Diana Elia, additionally refer to traditional techniques, such as the floor plan and the magic circle. Since ancient times, these techniques have mediated different dimensions of space, similar to *areal space*, thus establishing a bridge between history and latest technology.

1. What is an *Areal Space*?

The *areal space* is based on the interplay between spatial perception and spatial projection through technology. This is a concept that I developed during my research on AR.[10] It became increasingly clear throughout this research, how AR, perhaps more than related procedures, reveals fundamental aspects of XR. The

9 For a more in-depth analysis of the Guardian, see Stephan Günzel's contribution to this volume.

10 Cf. Manuel van der Veen (2025): Augmented Reality, Leiden: Brill/Fink, https://doi.org/10.30965/9783846769409, especially Ch. 2 »Kollision von Bild und Umgebung. Die Errichtung des Areals«.

following will show that the above-mentioned varieties of VR are only possible because they are based on augmentation processes as well. If one looks at conventional understandings of AR and VR, both have of course different tasks: AR embeds an image object into the existing surroundings, while VR replaces the existing surroundings with a virtual world. But if one thinks about what enables AR and VR, it is noticeable that they share the same properties. In VR, the image stays at its place even if one turns the head or if one moves through the virtual world, and so does AR. One can circle a digital object with a tablet in hands, while the object stays at its position in the actual surroundings. This is possible because the devices know where they are as a result of the 3D model of their surroundings. This is the first condition that AR and VR have in common: for both tasks – the creation of a world or the embedment of a single image object – the technology has to perceive its own position in space in relation to the actual surroundings. A condition that is more obvious with AR than with VR. The second condition they have in common is that they both project and thereby create space, which is in turn more obvious for VR and at least confusing for AR. Nevertheless, there exist a number of investigations in art that highlight precisely these conditions.

In the AR experience *Gold Environment* by artist Chiara Passa, the actual surroundings are gradually covered with golden plates, which then serve as a breeding ground for fictional plants (fig. 2a). Passa makes what happens in the background of every AR application visible. With AR, the surroundings are perceived by a LiDAR scanner and/or plane detection and subsequently are converted into a spatial model, which is then projected back onto the surroundings through alignment. Thus, the golden surfaces simultaneously materialise the alignment and creation of the spatial model. They demonstrate how the representation of space superimposes the represented space to produce a double ground. They also demonstrate how the detailed and crowded environment is transformed into an abstract sculptural relief. In conventional AR applications, the surroundings can only be seen through the real-time video image, which obscures the fact that they are also modeled and shaped in the background of AR. An artificial ground that completely overlays the actual one. In this way, the exterior space in *Gold Environment* is increasingly covered by this double ground until only a golden environment remains visible, thus gradually staging the transition from AR to VR.

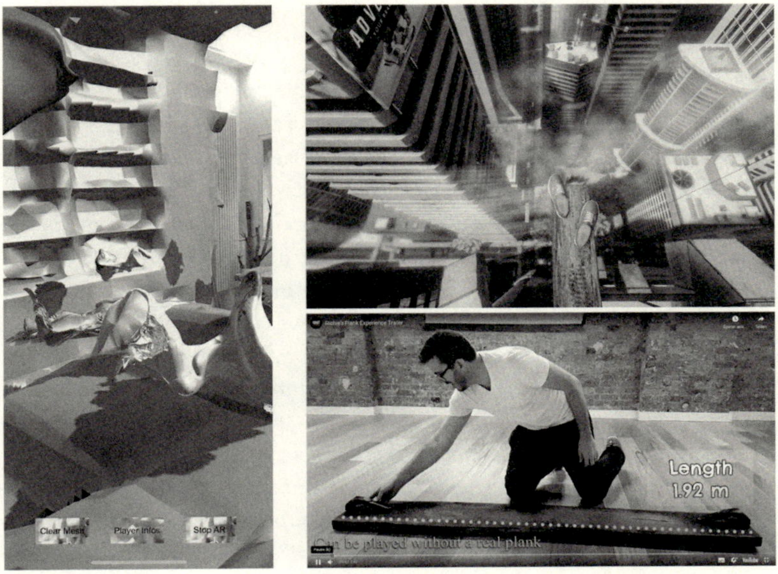

fig. 2a-c: left: Chiara Passa: Gold Environment, 2022, AR; right: Screenshots from video: Richie's Plank Experience Trailer

However, this 3D model of the actual space is always visible in AR as the negative space surrounding the digital objects. This becomes particularly obvious when shifts occur between the spatial model and the surroundings, causing the digital objects to shake, revealing their uncertain positioning between different areas. The superimposition of the represented space by the spatial representation creates a floating ground that can be congruent or slightly offset. In the latter case, corners turn into virtual curves, objects seem to float or pass through the solid walls. Firstly, this is important because if the 3D model does not appear to be completely congruent with the actual space, conflicts become apparent that fundamentally determine AR. Thereby, it reveals the differences that constitute the aesthetically relevant distance between the multiple dimensions. Secondly, this is important since the double ground is the foundation of the *areal space*.

1.1 Theory of the Double Ground

Before discussing different variants of the double ground in VR, this section first considers the double ground in theory. According to Nancy, the ground is of relevance to technology in general. More precisely, he uses the ambiguity of the term foundation (*Grund)*[11] to specify the Aristotelian concepts of *phýsis* and *techné*. The ambiguity of foundation is thereby understood »in the sense of the ground and in the sense of the principle of reason [i.e. *Grund, im Sinne des Bodens und im Sinne des Satzes vom Grund*].«[12] In English, this connection of ground and reason is part of the term ›explanation‹. Literally, to explain means to expose a plane. The roots lie in *ex-* for ›out‹ and in *plānus* for ›plane‹ or in *plicare* for ›to unfold‹, which therefore means ›to make something flat‹, i.e. to flatten a complex meaning. Hence, according to Nancy, *phýsis* (nature) could be understood as something that is a foundation for itself. This means that it grows from its own ground and finds its purpose in itself. *Techné,* on the other hand, is the area in which something is produced for which the foundation is not already given, as the foundation still must be found or invented. This is also where art and technology meet; in both fields, objects are created for which there is no foundation, no reason, no ground. All the various meanings of ›Grund‹ are important to subsequently identify art as a productive technique of foundation. Nancy also typifies art as production of presence by combining it with the act of »taking place«[13]. With other words, ›taking place‹ as the establishment of a foundation finds a place for something for which there is no ground.

In his considerations, Nancy mentions Aristotle's example for technique, i.e. the bed, and adds that for the bed there is no reason. Of course, there is a reason and a need for the bed, but people could just as easily sleep on the ground. For those who do not want to sleep on the ground, one must create

---

11 Cf. Gottfried Boehm (2012): »Der Grund. Über das ikonische Kontinuum«, in: Id./Matteo Burioni (eds.), Der Grund. Das Feld des Sichtbaren, Munich: Wilhelm Fink Verlag, pp. 29–91.

12 Cf. Jean-Luc Nancy (2008): »Destruktion als Erinnerung der Struktion oder Techné«, in: https://ifm.rub.de (27.11.2008). Online: https://ifm.rub.de/2008/11/1012-jean-luc-nancy-zu-gast-im-bkm/, 00:38:10 (last access: 10.10.2025), translation by author.

13 ›Taking place‹ as *stattfinden/eine Stätte finden*, which means ›to find a place‹. Cf. Jean-Luc Nancy (2013): »The Technique of the Present«, in: Documents of Contemporary Art 4/2013, pp. 104–115.

an artificial ground with the bed.[14] There is only a minimal difference between this foundation and the ground, for it is simply an elevated ground. Consequently, the bed is a duplication of the ground, cut-out and raised. In this respect, for Chiara Passa's *Gold Environment* the double ground of AR could also be understood as a technical foundation, just like the bed. Since digital things and spaces do not actually have a specific place, an artificial ground is laid out in situ with the 3D model on which they can be embedded in order to take place. This is one of the reasons why these technologies raise questions about what it means today for something to exist, to happen, to be present, to take place.

In VR, a projected space dominates, covering the entire field of vision, but what remains palpable is the double ground on which the user stands. Due to the encompassing virtual world, the perception of the outside world via VR technology usually quite literally fades into the background. The dominance of the projected space obscures the fact that for the enclosure of a virtual world, the surrounding space is also measured by cameras and sensors. With regards to the double ground, *Richie's Plank Experience* is one example that comes to mind. The popular app is based on a rudimentary approach, as the alignment of the spatial model is done manually rather than technically. Despite, or perhaps because of its simplicity, one could say that this game provides the scene of the *areal space*, or at least a scene in which shifts occur between the *areal space* of AR and VR. In this VR game, users take a lift to the 80th floor of a building, where they step out of the lift onto a plank at a dizzying height (fig. 2b). Even though they are still on the ground floor in reality, sweat is dripping from their foreheads and sticking on their hands and feet, as the fear of virtual heights has an actual effect. The effect is intensified with every cautious step on the plank. The app can be played only using a headset, but the developer's video shows how an actual plank can be added. A piece of wood that is 192 cm long, 18 cm wide and 4 cm high (fig. 2c). The plank provides a small rectangular area that, in Nancy's words, creates »a slight, faint, suspended reality«. And this wobbly, four centimeters raised ground makes it possible to experience the entire difference of 80 floors in the virtual realm. Users only slightly leave the ground of reality and yet find themselves in a different world.

---

14 This artificial ground shifts the underlying causes rather than resolving them. People sleep more comfortably on the bed, but they also become more sensitive and need even softer beds or training equipment. Cf. J. L. Nancy: Destruktion als Erinnerung der Struktion oder Techné.

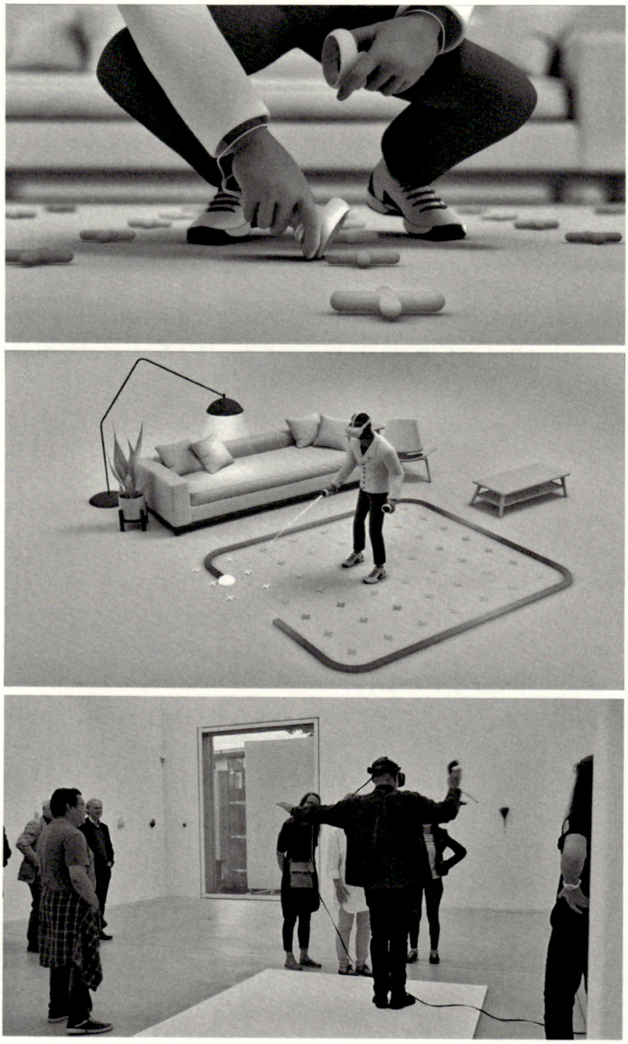

fig. 3 a-c: above: Screenshots from video »Quest 2 Setting Up Guardian«; below: Karin Sander: Virtual Reality Visit, 2018, installation view: Karin Sander, Kunst Museum Winterthur

The latter example shows how the establishment of a double ground can vary. Spatial recognition is present in all technologies of virtuality, whether through indexical recognition, which uses lasers to create a complex and detailed 3D model, through surface recognition, using position determination and contrast ratios, or through the manual setup of a virtual space in a specific room. Even the most rudimentary VR is based on the perception of the surroundings. One argument for this is the manual setup process for VR headsets, which is the alignment of the floor level in the virtual space with the floor level in the external space. To do this, the actual floor has to be touched with the controller (fig. 3a), which adjusts the floor level in VR to that of the outside space. This double floor or double ground reduces the surrounding space to a flat plane (fig. 3b), but even here a contact between the levels remains – triggered by touch. However, newer headsets scan a more complex relief of the surroundings, similar to AR.

1.2 The Three Aspects of the *Areal Space*

Based on these considerations, the condition of the possibility of XR is not the superimposition of digital objects onto the actual surroundings or the creation of a comprehensive virtual world. Instead, it is primarily the alignment of a spatial model in situ and in real time. Therefore, AR and VR do not necessarily have to be interactive, but from a technical perspective, spatial perception must interact with spatial projection.

Generally speaking, it is about the collision of perception and projection, i.e. how the surroundings are perceived by the technology and how the technology then projects something onto the surroundings. Only a perception or projection through technology alone is never decisive and the interplay of realities takes place between them, whether visible or invisible. The concept of *areal space* is a way of describing a specificity of XR to set them apart from other technologies such as computer games, spatial projections or panoramas. It is a cumulative definition with three aspects, and when all three occur together, one can speak comprehensively of AR, VR, or MR. Hence, the *areal space* also allows to identify (historical) relatives that might only share one or two aspects of it. Even so, it can be useful for a deeper analysis of the latest technology.

The first aspect is that there must be projective and perceptive elements for XR. This is crucial as it results in two opposing directions. On the one hand, data is received from the surroundings and, on the other hand, data is projected from the data storage (fig. 4a). Due to the different directions and sources, their encounter initially leads to collision, creating contradictions and conflicts. This condition is necessary but as a single feature not tied to digital technologies, as it touches on both machine vision techniques and classical theories of human perception.[15] Mirrors, glass panes and projectors can also be used to implement perceptive and projective elements. For example, the phantasmagoria (fig. 4b). In *Peppers Ghost* the reception of the stage collides with the projection of a figure underneath the stage on a glass pane.[16] Heterogeneous logics collide, and their differences are part of the mixed experience, such as those between supernatural and physical bodies or digital and everyday objects. With the first aspect, it can be asked: what collides and in what intensity? For example, virtual worlds and living rooms collide in VR, and while the projective elements are encompassing, the perceived elements can be rudimentary. The collision explains why these technologies are frequently promised as new realities but are rather experienced as conflicts that reach to the point of motion sickness. Thus, the first aspect is that for XR there have to be perceptive and projective elements, which collide on an *interface of the areal space* (headset/screen/glass pane).

In order to get closer to the specific characteristics of XRs, the perceptive and projective elements must secondly be related to each other, i.e. the *double ground of areal space*.[17] This can be done manually or technically. Thus, the phantasmagoria already fulfills two aspects of *areal space*, as it is also a double floor that raises the ghost one level higher than it actually is. In XR, cameras and sensors perceive the immediate surroundings and create a 3D model of it – either flattened or finely modelled – in order to subsequently calculate this representation of the surroundings point by point with the represented surroundings. Only after establishing a relationship, a double ground, it is possi-

15 See my considerations on Arthur Schopenhauer's receptive-projective theory of perception in Manuel van der Veen (2022): »Zur Technik-Ästhetik der Mixed Reality. Über das Verhältnis von Redundanz und Augmentation«, in: Lars C. Grabbe/Oliver Ruf (eds.), Technik-Ästhetik. Zur materialen und rezeptiven Systematisierung techno-ästhetischer Realität, Bielefeld: transcript, pp. 203–222, https://doi.org/10.14361/9783839456361-011.

16 Cf. Noam M. Elcott (2016): »The Phantasmagoric Dispositif. An Assembly of Bodies and Images in Real Time and Space«, in: Grey Room 62, pp. 43–71.

17 Therefore, we have to talk about the resulting images as adaptive ones. See Matthias Bruhn/Kathrin Friedrich/Moritz Queisner (2021): »Adaptive Images. Challenges for Visual Studies and Media Theory«, in: International Journal for Digital Art History 1, pp. 140–53, doi.org/10.11588/dah.2021.E1.83933.

ble to superimpose a virtual world onto the surroundings or embed individual digital objects within it. Therefore, the second aspect allows to ask: How is the relationship constituted? Is the spatial model detailed or abstract? Are differences provoked or is the aim to achieve convergence?

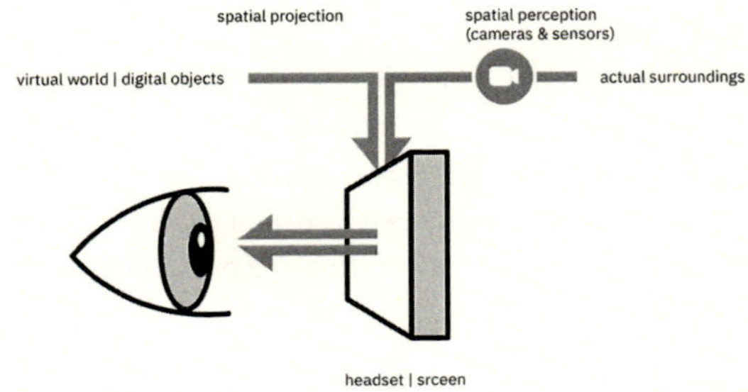

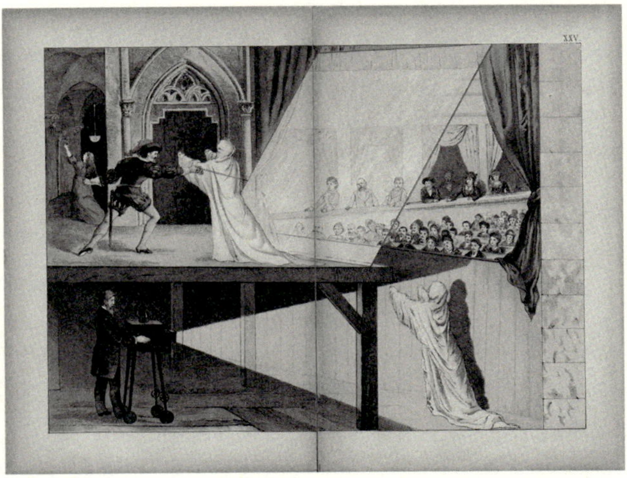

fig. 4 a&b: above: illustration of the processes and components making up XR (modified by the author); below: Illustration of Pepper's Ghost from Die Physik in Bildern, color plate XXV

The mobility of the areal space is the third aspect, since the double ground is not manifest and the anchoring is only valid for a certain period of time until the *areal space* is carried away.[18] The digital model of the surroundings must be technically created and constantly updated for its mobility. The »uncertainty«[19] of this connection arises from the change of the surroundings and the position of the device within them, i.e. from the simultaneous localisation and mapping (SLAM). After all, the headset/screen needs a map for localisation, and in order to create a map, it has to localise itself. Since both the headset and the actual objects in the surroundings are in motion, this world modelling generates probability values and thus a constantly »fragmentary model«[20]. *Areal space* is where the device is carried to, where the gaze and the sensors are directed to, and it extends as far as they reach. Although it can be defined as an unreal place, it is not independent of a physical place. The third aspect therefore raises the question of how mobile the *areal space* is and shows why contemporary artists can use VR not only as a portal to unfamiliar worlds, but also to create site-specific works.

The three aspects were derived from the essential components of XRs: 1. the device (headset/screen/glass pane), 2. its relation to the surroundings, and 3. the interdependence between the two. These aspects define the foundation of the *areal space*, which takes place before it is occupied by projected objects or entire virtual worlds. Therefore, these aspects are also suitable for distinguishing between different XRs, so that various works of contemporary art can now be considered.

2. The *Areal Space* of Art with VR

The *areal space* is occasionally addressed directly within VRs, as in *Blooming Love* (2021) by Daniel Hengst. In this VR, the viewers find themselves in the middle of a bog, where new plants are constantly being generated and slowly passing

18 Regarding the mobility of AR see Nanna Verhoeff (2012): Mobile Screens. The Visual Regime of Navigation, Amsterdam: University Press, online: https://www.jstor.org/stable/j.ctt46mtwb (last access: 18.10.2025).

19 Florian Sprenger speaks of uncertain environments in connection with adaptive technologies that must be able to move automatically in their environment. Although a headset must be able to orient itself in the environment, movement remains the responsibility of the user. See Florian Sprenger (2019): Epistemologien des Umgebens. Zur Geschichte, Ökologie und Biopolitik künstlicher *environments*, Bielefeld: transcript, pp. 484–486.

20 Ibid.

by. For the experience, Daniel Hengst raises the virtual floor level so high that the viewer is stuck up to the neck in the bog. Raising the floor level creates a noticeable gap between the digital space and the exhibition space. In the work *Blooming Love*, the double floor becomes visible as it falls apart. Furthermore, there are several current VR experiences that also reflect the *areal space* in the exhibition. Therefore, the exhibited installations adhere the following pattern: a specific area in the exhibition is marked with lines, a platform, a carpet, soil, etc., where the VR experience then takes place. What is striking here is how the double ground repeatedly negotiates the difference between limited space and extended space.[21] Hence, a simple white platform, placed slightly elevated, can be set up to show a complete virtual exhibition, as in Karin Sander's exhibition *Karin Sander* (fig. 2c). In the following, two exhibitions that explore the double ground are examined: Sander's exhibition and *Drifting, Browsing, Cruising* by Polyviou and Elia. Both address the specificity of the *areal space* while different spatial models collide. These embrace drawing and architecture.

2.1 Real Time and Real Scale Floor Plans

In the following, the two spaces – the space of a drawing and the built space – are the focus of attention. Firstly, because AR and VR generally blur the distinction between 2D and 3D. Secondly, because drawing a space is a fundamental part of VR, as the Guardian example shows (fig. 3b). And finally, because the difference between the spaces highlights the collision and the conflicts it causes. In the anthology *Bildhafte Räume, begehbare Bilder* Kassandra Nakas and Philipp Reinfeld focus on the discrepancy that has existed since the Renaissance between two-dimensional architectural drafts and three-dimensional built spaces. Instead of nourishing this discrepancy, the various contributions consider draft and construction together in order to »examine real-time and real-scale design in VR.«[22] In this regard, Jan Philipp Drude refers to the cathe-

21 In research, attempts are being made, for example, to enable people to truly walk through an infinite virtual space by means of redirected walking, which involves walking along curves in real space. See Frank Steinicke et al. (2010): »Estimation of Detection Thresholds for Redirected Walking Techniques«, in: IEEE Transactions on Visualization and Computer Graphics 16/1, pp. 17–27, here: p. 25, https://doi.org/10.30965/9783846767238.

22 Kassandra Nakas/Philipp Reinfeld (2022): »Bildhafte Räume und begehbare Bilder. Zur Einführung«, in: Id. (eds.), Bildhafte Räume, begehbare Bilder, Leiden: Brill/Fink, pp. 1–7, here: p. 1, https://doi.org/10.30965/9783846767238, translation by author.

dral construction site around 1200: »Drawings on a scale of one to one were scratched directly into the floors and walls of the cathedral construction site, as can be seen, for example, in the tracing floor [*Rissboden*] of York Minster.«[23] Drawing the floor plan on site creates a parallel to the Guardian, which consists of ›walls‹ built from lines on the double floor. This parallel highlights the heterogeneous logics implemented within the technology and its historical connections. However, this connection goes even further: it places VR between draft and built space, emphasises it as a technology of scaling and demonstrates its potential as a site-specific medium.

Karin Sander, *Karin Sander*, 2018

Of course, double floors also exist in the field of art, without relying on XR. For this it could be referred to Karin Sander's work *Floor (Boden)* in the 2018 exhibition *Karin Sander* at the Kunstmuseum Winterthur (fig. 5a). The floating reality of the floor makes the viewer's steps cautious and the surrounding emptiness swaying. The viewer's attention is thereby divided between the unknown of a protruding platform and the all-too-familiar accessible plane. Standing on the raised floor, they are aware that this is not everyday life. The artist thus declares the museum space to be an artificial space, emphasised by the elevated stage.

To examine more closely how Sander connects different spaces with her works, the exhibition structure must first be briefly presented here on the basis of the floor plan (fig. 5c). Sander shows various series of works in her exhibition, which are separated by small rooms. As can be seen in the floor plan, the separate rooms are connected transversely by the work *Floor (Boden)* (the blue lines in fig. 5c) and longitudinally by a partially constructed brick wall (red lines in fig. 5c, visible in fig. 5b). Furthermore, the whole exhibition is framed by an outline consisting of Sander's *Kitchen Pieces* (red and blue dots in fig. 5c, on the left wall in fig. 5b). This arrangement is crucial because it takes place in two different exhibitions, one in physical space (blue elements) and one in VR (red elements). Sander places the VR exhibition *Virtual Reality Visit* in the museum, for which the entire annex of the museum was recreated in high resolution. Consequently, she was able to curate two exhibitions. However, the museum's annex was not simply adopted but rebuilt for VR. The first modification involves replacing the double floor of *Floor (Boden)* with that of VR. Just

23 Jan Philipp Drude (2022): »Vom Reißbrett in die Virtuelle Realität: Wie sich unsere Entwurfsperspektive verändert«, in: K. Nakas/P. Reinfeld: Bildhafte Räume, begehbare Bilder, pp. 11–26, here: p. 12, translation by author.

as *Floor (Boden)* connects the individual rooms, VR connects the digital model with the exhibition space. In fact, the relation between the two exhibitions is structured both by a separation and by a connection. On the one hand, Sander had the dividing walls of the exhibition space between the rooms 12, 13 and 14 removed in VR to underscore the connecting outline of the *Kitchen Pieces* (fig. 5b). On the other hand, a separation is inserted into the exhibition space, as the bricks from the ›removed‹ walls now block the passage in the digital space that *Floor (Boden)* had just bridged in the exhibition space. In the exhibitions, two double floors are laid, negotiating the relationship between separation and connection.

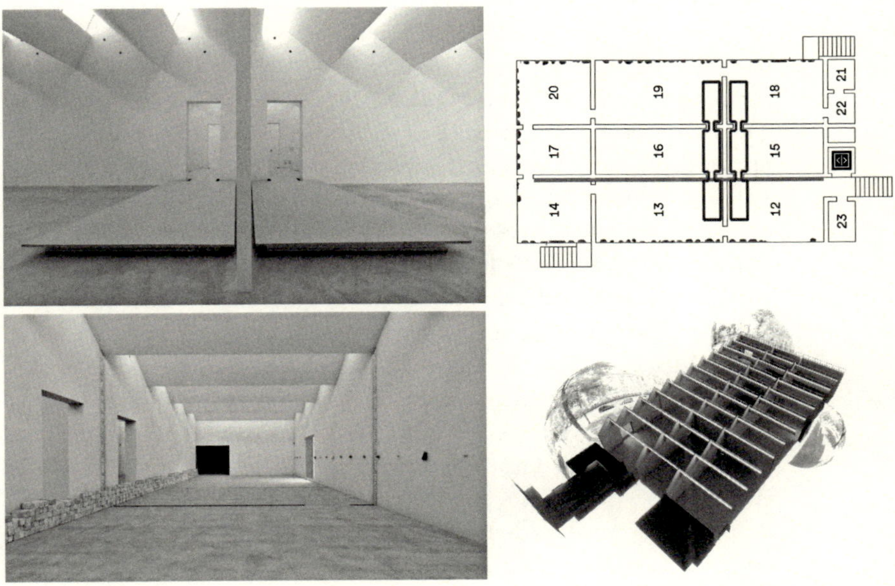

fig.5 a-d: tl: Karin Sander: Floor (Boden), 2018, exhibition view: Karin Sander, Kunst Museum Winterthur; bl: VR view of Karin Sander: Virtual Reality Visit, Kunst Museum Winterthur with virtually removed walls, 2018, virtual reality technic, 3-D visualization; tr: floorplan of the exhibition; br: 3D model of Karin Sander's VR exhibition: Virtual Reality Visit

The entanglements between the different spaces thus become crucial. In the VR exhibition space, for example, there is a centre that cannot be entered (room 15 and 16). A circumstance that, on the representational level, repeats the characteristic of VR of being equally inaccessible from the outside. Moreover, in VR, the *Kitchen Pieces* create an outline, a specific area that surrounds the viewers. An area that is physically separated from the outside world in the exhibition space, as Sander blocks the view through the windows with large paintings, which are called *Patina Pictures (Gebrauchsbilder)*, in order to create a hermetically sealed exhibition space. This enclosed space is then reopened in the digital model by using 360° photographs of the outside space mapped onto spheres (fig. 5d).

The 3D model of the space in VR usually remains invisible. In this case, however, the transparency of the model breaks through the walls of the spatially structured architecture, and the intended hermetic nature of the exhibition is reopened in VR. This clearly shows how the artist has already anticipated the VR exhibition within the exhibition in physical space. Sander thus simultaneously negotiates the isolation of the museum and its architectural requirements for an exhibition with a specific feature of hermetic VR. A specific feature that directly addresses the everyday use of VR, as the Quest 2 tutorial states: »First, let's find a play area in a well-lit room and clear as much space as possible. Remove any obstacles which you might collide with while you're in your virtual environment.«[24] In Sander's work, the emptiness of a white cube[25] appears in analogy to the cleared-out area of VR, which is demarcated by the Guardian. VR is therefore not just about replacing reality with a virtual world. Rather, VR has an impact on reality. After all, like the grid-like floor level, reality must provide a smooth, resistance-free, and bare surface – this is an actual reduction of reality. Consequently, the physical space must be truly modified for the sake of VR, as it can only be projected where the surroundings have been cleared.[26]

Conversely, Sander's confrontation makes the White Cube appear as a projection surface for hermetic worlds. She clears away entire walls of the existing architecture in the process. As a result of clearing the space, lines are exposed

---

24 Meta Quest (2021): »Quest 2 Setting Up Guardian«, in: youtube.com. Online: https://www.youtube.com/watch?v=GojevL05Avw, 00:00:27 (last access: 09.12.2025).

25 Cf. Brian O'Doherty (1986 [1976]): Inside the White Cube. The Ideology of the Gallery Space, Santa Monica/San Francisco: The Lapis Press.

26 This is reminiscent of the long and empty corridor of central perspective. See Joseph Masheck (1991): »Alberti's ›Window‹. Art-Historiographic Notes on an Antimodernist Misprision«, in: Art Journal 50/1, pp. 35–41, here: p. 36.

on the ground at the same time. The tearing down of the partition walls reveals the floor plan, because lines can now be seen on the floor, as they are already marked on the floor plan. With the double ground, Sander emphasises the possibility of re-construction, beyond a rigid replica, by placing VR between the draft and the final construction, between a drawing and a built space. In VR, Sander reverses the construction process, removes the resistance of the structuring architecture, and thus brings about the augmentation of reality together with its reduction. To pursue these questions further, the next example introduces another technique that, like the floor plan or tracing floor (*Rissboden*), connects different dimensions and is directly related to virtual worlds.

2.2 Magic Circles as ›Imaginable Architectural Devices‹

In addition to the floor plan, the magic circle is considered here, as it also involves marking an area on the floor with lines that create a space, in this case a supernatural one. Furthermore, the magic circle has a tradition as a concept in game studies and was also transferred to computer games. The latter transfer was made by Jesper Juul in his 2005 dissertation *Half-Real*.[27] This concept is therefore particularly suited to mediate between the tradition of virtual spaces and the latest technology. A magic circle is the demarcation of a physical area, usually by carving lines into the ground, thereby establishing the boundaries within which an exchange between this world and the other world can take place. While the act of drawing lines is directly parallel to the establishment of the *areal space* by the Guardian in VR, its function varies in magical, playful and virtual practices. Katie Salen and Eric Zimmermann state the following about the magic circle: »The term is used here as shorthand for the idea of a special place in time and space created by a game.... As a closed circle, the space it circumscribes is enclosed and separate from the real world.... In a very basic sense, the magic circle is where the game takes place«[28]. As long as someone

27 Cf. Jesper Juul (2005): Half-Real. Video Games between Real Rules and Fictional Worlds, Cambridge, MA/London: The MIT Press. The magic circle was already discussed in *Les rites de passage* by religious ethnologist Arnold van Gennep, and the idea of understanding a game by means of the magic circle goes back to Johann Huizinga's *Homo ludens*. Cf. Stephan Günzel (2010): »Der reine Raum des Spiels. Zur Kritik des Magic Circle«, in: Mathias Fuchs/Ernst Strouhal (eds.), Das Spiel und seine Grenzen. Passagen des Spiels II, Vienna: Springer, pp. 189–202, here: p. 196, https://doi.org/10.1007/978-3-7091-0085-1_12.

28 Here quoted after J. Juul: Half-Real, p. 164.

is inside this area, the rules of the game apply. Not every game has such clear demarcations as, for example, the field for the children's game of hopscotch (*Himmel und Hölle*). However, according to Juul, the boundary is even more distinct in a computer game, as it »only takes place on the screen«[29]. This makes the model particularly interesting in comparison to the *areal space* of VR.

Juul's motivation to use the magic circle for computer games was the attempt to reconcile the real elements of a game (the ludological category of rules) with the fictional ones (visual and narratological elements). The accompanying illustration (fig. 6d) can be summarised as follows: »The space of a game is part of the world in which it is played, but the space of a fiction is outside the world from which it is created.«[30] Stephan Günzel emphasises that Juul's magic circle no longer represents a boundary in the world, but rather reduplicates the boundary between fiction and reality outside the ›real world‹. Günzel further criticises that the rules in a computer game differ from the rules of a conventional game. After all, one cannot choose to abide by the rules of the computer game: »The game is simulated physics at every level – with the exception of the area that Juul calls fiction.«[31] Consequently, this argument has an impact on the permeability of the magic circle. There is only a transition from the fictional world to the game space, as the former can be used to anticipate the latter. The skyscraper and the plank in *Richie's Plank Experience*, for example, imply that one can probably jump. However, there is no way to leave the system of rules, the game space to immerse oneself in the fiction outside it. Thus, one cannot simply climb down the facade because that is not part of the prescribed interactions in *Richie's Plank Experience*. The magic circle is therefore not a suitable model for representing the relationships between real and fictional elements in computer games. The model also fails to highlight the specificity of computer games, as this specificity lies neither in the narrative (fiction) nor in the rules (reality), but rather – according to Günzel in his text *Der reine Raum des Spiels* – in the fact that »its reception [that of computer games] does not require any external reference, but is based solely on action relying on spatial structures. [...] Contrary to Juul, it can be said that computer games can be distinguished from all other types of games precisely by the absence of a magic circle.«[32]

29 Ibid., p. 165.

30 Ibid., p. 164.

31 S. Günzel: Der reine Raum des Spiels, p. 200, translation by author.

32 Ibid., p. 200f., translation by author.

But what about the magic circle when it relates to VR? After all, the ›fictional world‹ is not projected onto the outside of the ›real world‹, but rather superimposed on it. As a result, it could be asked: Does the simulated physics overlap with the real physics? Can they influence each other? Even if there is no escape from simulated physics in VR either, the concept of the magic circle from game studies still raises questions that can be used to further differentiate the area of the virtual.

Theodoulos Polyviou and Eleni Diana Elia, *Drifting, Browsing, Cruising*, 2021

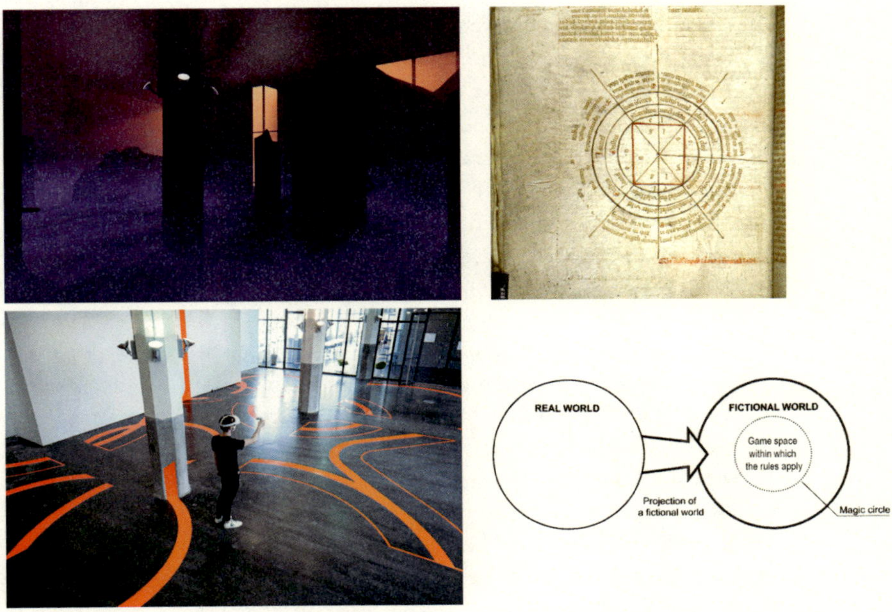

fig. 6 a-d: tl: Theodoulos Polyviou and Eleni Diana Elia: Drifting, Browsing, Cruising, 2021, ZKM Karlsruhe, VR view; bl: installation view; tr: diagram of magic circles for the conjurations of airy spirits; br: Jesper Juul's diagram of the magic circle for computer games (modified by the author)

Drifting, Browsing, Cruising by Polyviou and Elia was on display at the ZKM in Karlsruhe in 2021. Instead of emptying a full room as Sander did, an empty room is virtually filled up with obstacles here. A floor plan consisting of thick orange lines, is therefore not derived from the existing architecture, but added to it. Subsequently, the brightly glowing lines on the floor of the exhibition are erected in VR as impenetrable ruin walls. In this respect, the artwork produces a categorical transformation from a mundane exhibition space to a sacred ambience. While the exhibition space allows for an overview due to its emptiness and clear markings, the projected space is of unpredictable density. In the VR, diffuse light falls on purple mist, while columns, struts, railings and walls create resistance as almost black surfaces and block lines of sight. In contrast to the exhibition space, it is difficult to orientate oneself in the digital mist.

Polyviou and Elia's work opens up institution-critical perspectives[33] and addresses various modes of encounter in digital, physical and virtual space. The encounter between different entities, the lines on the ground and the virtually augmented architecture should be interpreted here beyond the floor plan via the supernatural magic circle. Polyviou refers directly to the considerations of the historian of religion Andrea Franchetto. He describes the magic circle as an »imaginable architectural device«[34] through which ritual elements are expressed in architectural objects. For this purpose, the drawing is scratched into the ground, forming the floor plan of a cosmological topography (fig. 6c). This creates a paradoxical space in which two areas communicate because the imagined boundaries are materialised.

> »In other terms, the lines on the ground become material anchors that substantiate the presence of the internal image boundaries. [A] blended space, which is the fusion of the mental and the material domain. It is thus augmented materiality – spaces ›called up‹ in order to actualize a new set of desires and power relations […]«[35]

The drawing is architectural because the lines are protective walls – they enclose, demarcate and provide orientation for the control of space as well as for the movement of practitioners. The dramatic effect of a magic circle consists

---

33 Cf. Annette Urban (2023): »Virtual Spaces for Transformative Encounters and Vast Reciprocity – An Interview with Theodoulos Polyviou and Jazmina Figueroa«, in: Lívia Nolasco-Rózsás/Marianne Schädler (eds.), Beyond Matter, Within Space. Curatorial and Art Mediation Techniques on the Verge of Virtual Reality, Berlin: Hatje Cantz, pp. 429–443, here: p. 440.

34 Cf. Andrea Franchetto (2020): »Imaginal Architectural Devices and the Ritual Space of Medieval Necromancy«, in: Endeavour 44/4, pp. 1–16.

35 Ibid., p. 6.

in the fact that it is not allowed to have its lines broken, otherwise the different dimensions would merge outside the demarcated area. Polyviou and Elia's circles are interrupted many times. After all, the aim is to provoke an encounter between different spheres and people.

In order to position the magic circle of the Middle Ages, computer games, and VR in relation to each other, certain shifts must be highlighted. Firstly, it is crucial to note that the literal demarcation of an area rarely applies to games in general and does not apply to computer games at all. In the set-up of VR, however, this very process is repeated: an area is marked as a drawing in which the virtual world takes place, and if this area is crossed by the user, the virtual world dissolves. However, since the boundary is only drawn in VR, its status is at least problematic.

Secondly, it is therefore important that the conventional magic circle may establish multiple boundaries: an inner circle that protects the magicians and an outer circle in which the supernatural beings appear and are held captive. It is this variety of boundaries that is also addressed in contemporary art with VR. Some protect users and others exclude people outside of VR. And then there are those that even include outsiders to a certain extent in the *areal space*, although they are not wearing a headset. Even if, contrary to Juul's illustration, fiction in VR is projected directly onto reality, the two do not automatically overlap. When a person enters the area marked by Polyviou and Elia in the exhibition space, they are not part of the VR but they enter the *areal space*. The person wearing the headset cannot see these ›intruders,‹ however, they influence the experience by infiltrating the cleared area. VR is also not completely hidden from outsiders, as they can observe how the person wearing the headset respects and avoids the virtual boundaries. So, as with the magic circle, it does matter whether one is inside the *areal space* or not.

The third point concerns the rules of the game, which, according to Günzel, appear as simulated physics in computer games. One cannot act against the simulated physics in VR either. However, since one's own body is involved in VR and the simulated physics overlap with the real ones, there is room for maneuver (*Spielraum*) in which a kind of creative waywardness (*Eigensinn*) is possible. This means that the users can perform gestures that the developers could not have foreseen. They can try to lean out of the world, exploit dislocations between the model and the surroundings as well as break through collision boxes that apply to the avatars but not to their own bodies.

The magic circle emphasises the interplay between imagination and physics, which is becoming increasingly relevant for the exploration of contemporary virtual art. Even though simulated physics cannot be violated in a virtual world, the *areal space* opens up a »system of relations between different elements that can be found in society, and which strategically manipulates power and knowledge in order to orient, direct, and control those elements.«[36] The *areal space* is a structured space that also has a structuring effect on the actual surroundings outside of VR.

Conclusion

This chapter presented the *areal space* as an overarching concept for interpreting XRs. The concept identifies the technical specificity of the procedures – to be both space-perceiving and space-projecting – beyond ontological dichotomies such as ›virtual‹ versus ›real‹. Therefore, this concept can be used to differentiate more precisely between the technologies of VR, AR, and MR, as well as between different VRs. Regarding the first aspect of the *areal space*, i.e. the interface, it can be said that in Sander's VR, a full and structured space is perceived, and an empty one is projected, while both remain comparable due to their similarities. However, there is no direct collision, as the two exhibitions are experienced independently of each other: one on foot and one via controllers. In contrast, Polyviou and Elia's VR builds on strong visual differences to the exhibition space, where the perceived and projected space provokes actual collisions due to their superimposition.

In relation to the double floor, it is crucial that in Sander's VR, the existing space calls for an onsite renovation in VR, which reveals the drawing of the floor plan. *Drifting, Browsing, Cruising*, instead, was presented as a space in VR that causes a drawing in the exhibition space. This inversion shows how the effects of the double floor can occur in both directions. Nevertheless, in both works discussed, the double floor creates a site-specific area by superimposing the critique of the exhibition space onto its usage. Once through its modification and once through its sacralisation. Yet, the double floor does not necessarily result in site specificity; as in *Richie's Plank Experience*, it can also intensify the affective experience, or, as in Daniel Hengst's VR, mark the distance between worlds, or simply define the field of experience. Of course, there are also VRs, which, apart from the technically necessary spatial references, continue to focus on the hermeticism of new worlds.

36 Ibid., p. 4.

Furthermore, it could be mentioned that despite the fact that Sander's VR and Polyviou and Elia's VR are site-specific, the mobility of the *areal space* varies significantly. This is because the model in Sander's VR is exclusively tied to a specific exhibition. *Drifting, Browsing, Cruising*, by contrast, is more mobile, as the orange line, i.e. the imaginable architecture, can also be placed in a different location and because the entire digital space is already accessible as a detached online experience, which adds another layer to the double ground.[37]

Finally, alongside the potential for differentiation, it is the theoretical scope of the concept of *areal space* – its theoretical compatibility, not least for art historical questions – that may be of interest to XR research. The aforementioned floor plan or tracing floor (*Rissboden*) supplemented the concept with the tensions between 2D and 3D, draft and built space, which are also fundamental to XR. In the same way, the ›imaginable architectural device‹ extended the spatial projection and the spatial perception of the *areal space* into a relationship between imagination and materialisation. The drawing ties the imaginable to a physical space and VR back to the magical origins, on which AR is based: the magic lantern and the phantasmagoria. The virtual thus refers to a dimension of the ghostly and spiritual, which would be lost under a solely computer-technical classification. However, the comparison that the floor plan and the magic circle open up between everyday life, computer games, VR, and art, could not be played out here exhaustively. Rather, the aim was to show how technical, (art-)historical and theoretical considerations converge through the concept of *areal space* in order to approach VR as a virtual phenomenon.

Bibliography

Boehm, Gottfried (2012): »Der Grund. Über das ikonische Kontinuum«, in: Id./ Matteo Burioni (eds.), Der Grund. Das Feld des Sichtbaren, Munich: Wilhelm Fink Verlag, p. 29–91.

Bruhn, Matthias/Kathrin Friedrich/Moritz Queisner (2021): »Adaptive Images. Challenges for Visual Studies and Media Theory«, in:

37 See the online version of *Drifting, Browsing, Cruising*, which was subsequently made available for further exhibitions as *fantastic confabulations*, »an online platform for virtual, yet spatial exhibitions based on architectural motifs of the building, currently occupied by ZKM | Karlsruhe«, online: https://fantastic-confabulations.beyondmatter.eu/drifting-browsing-cruising/index.html (last access: 17.11.2025).

International Journal for Digital Art History 1, pp. 140–53. DOI: doi.org/10.11588/dah.2021.E1.83933

Chalmers, David J. (2022): Reality+. Virtual Worlds and the Problems of Philosophy, New York: W. W. Norton & Company.

Dresscher, Paulien/Nanna Verhoeff (2020): »XR. Crossing and Interfering Artistic Media Spaces«, in: Larissa Hjorth/Adriana de Souza e Silva/Klare Lanson (eds.), The Routledge Companion to Mobile Media Art, New York: Routledge, pp. 482–92.

Drude, Jan Philipp (2022): »Vom Reißbrett in die Virtuelle Realität: Wie sich unsere Entwurfsperspektive verändert«, in: Kassandra Nakas/Philipp Reinfeld (eds.), Bildhafte Räume, begehbare Bilder, Leiden: Brill/Fink, pp. 11–26.

Elcott, Noam M. (2016): »The Phantasmagoric Dispositif. An Assembly of Bodies and Images in Real Time and Space«, in: Grey Room 62, pp. 43–71.

Franchetto, Andrea (2020): »Imaginal Architectural Devices and the Ritual Space of Medieval Necromancy«, in: Endeavour 44/4, pp. 1–16.

Günzel, Stephan (2010): »Der reine Raum des Spiels. Zur Kritik des Magic Circle«, in: Mathias Fuchs/Ernst Strouhal (eds.), Das Spiel und seine Grenzen. Passagen des Spiels II, Vienna: Springer, pp. 189–202. DOI: https://doi.org/10.1007/978-3-7091-0085-1_12

Juul, Jesper (2005): Half-Real. Video Games between Real Rules and Fictional Worlds, Cambridge, MA/London: The MIT Press.

Masheck, Joseph (1991): »Alberti's ›Window‹. Art-Historiographic Notes on an Antimodernist Misprision«, in: Art Journal 50/1, pp. 35–41.

Meta Quest (2021): »Quest 2 Setting Up Guardian«, in: youtube.com. Online: https://www.youtube.com/watch?v=GojevLo5Avw (last access: 09.12.2025).

Milgram, Paul/Fumio Kishino (1994): »A Taxonomy of Mixed Reality Visual Displays«, in: IEICE Transactions on Information and Systems E77-D/12, pp. 1321–1329.

Nakas, Kassandra/Philipp Reinfeld (2022): »Bildhafte Räume und begehbare Bilder. Zur Einführung«, in: Id. (eds.), Bildhafte Räume, begehbare Bilder, Leiden: Brill/Fink, pp. 1–7. DOI: https://doi.org/10.30965/9783846767238

Nancy, Jean-Luc (2003 [1992]): Corpus. New York: Fordham University Press.

Nancy, Jean-Luc (2008): »Destruktion als Erinnerung der Struktion oder Techné«, in: https://ifm.rub.de (27.11.2008). Online: https://ifm.rub.de/2008/11/1012-jean-luc-nancy-zu-gast-im-bkm/ (last access: 10.10.2025).

Nancy, Jean-Luc (2013): »The Technique of the Present«, in: Documents of Contemporary Art 4/2013, pp. 104–115.

O'Doherty, Brian (1986 [1976]): Inside the White Cube. The Ideology of the Gallery Space, Santa Monica/San Francisco: The Lapis Press.

Reich, Julia/Annette Urban/Manuel van der Veen (2023): »passthrough. Von Portalen, Durchblicken und Übergängen zwischen den (virtuellen) Welten«, in: Kunstforum International 290, pp. 86–95.

Skarbez, Richard/Mary C. Whitton/Missie Smith (2021): »Revisiting Milgram and Kishino's Reality-Virtuality Continuum«, in: Frontiers in Virtual Reality 2, n.p.

Sprenger, Florian (2019): Epistemologien des Umgebens. Zur Geschichte, Ökologie und Biopolitik künstlicher *environments*, Bielefeld: transcript.

Steinicke, Frank et al. (2010): »Estimation of Detection Thresholds for Redirected Walking Techniques«, in: IEEE Transactions on Visualization and Computer Graphics 16/1, pp. 17–27.

Urban, Annette (2023): »Virtual Spaces for Transformative Encounters and Vast Reciprocity – An Interview with Theodoulos Polyviou and Jazmina Figueroa«, in: Lívia Nolasco-Rózsás/Marianne Schädler (eds.), Beyond Matter, Within Space. Curatorial and Art Mediation Techniques on the Verge of Virtual Reality, Berlin: Hatje Cantz, pp. 429–443.

van der Veen, Manuel (2022): »Zur Technik-Ästhetik der Mixed Reality. Über das Verhältnis von Redundanz und Augmentation«, in: Lars C. Grabbe/Oliver Ruf (eds.), Technik-Ästhetik. Zur materialen und rezeptiven Systematisierung techno-ästhetischer Realität, Bielefeld: transcript, pp. 203–222. DOI: https://doi.org/10.14361/9783839456361-011

van der Veen, Manuel (2025): Augmented Reality, Leiden: Brill/Fink. DOI: https://doi.org/10.30965/9783846769409

Verhoeff, Nanna (2012): Mobile Screens. The Visual Regime of Navigation, Amsterdam: University Press, online: https://www.jstor.org/stable/j.ctt46mtwb (last access: 18.10.2025).

Image Credits

fig. 1 a&b: a) Courtesy the artist, photo: Joy Lai; b) © Meta Platforms Inc., source: https://youtu.be/x17GeJ4cf9o?si=kj_rO_eZXZ4qrarQ

fig. 2 a-c: a) Courtesy the artist; b&c) © Toast Interactive, source: https://youtu.be/4M92kfnpg-k?si=IHtGEXdmN8ZD7Ijg

fig. 3 a-c: a&b) © Meta Platforms Inc., source: https://youtu.be/GojevLo5Avw?si= i14G83roBDFlwX_L (second image edited by the author); c) © Studio Karin Sander, photo: Andrea Rossetti

fig. 4 a&b: a) source: Vinod Baya u. Erik Sherman (2016): »Breakthroughs in Optics that are Reshaping Augmented Reality«, in: Technology Forecast. Augmented Reality 2, p. 3 (modified by the author); b) source: Die Physik in Bildern, Eßlingen: Verlag von J. F. Schreiber 1881, color plate XXV, online: https://leopard.tu-braunschweig.de/servlets/MCRFileNodeServlet/dbbs_derivate_00055328/images/00000104.jpg (last access: 17.10.2025)

fig. 5 a-d: a) © Studio Karin Sander, photo: Andrea Rossetti; b&d) © Karin Sander, VR simulation: Adam Kiryk, Professur Architektur und Kunst, ETH Zürich; c) source floorplan: https://www.kmw.ch/content/uploads/2018/09/Saalblatt_Karin-Sander_online.pdf (last access: 17.10.2025)

fig. 6 a-d: a&b) © Eleni Diana Elia/Theodoulos Polyviou, Zentrum für Kunst und Medien Karlsruhe, photo: Tanja Meissner; c) © British Library Board, Sloane 3854, fol. 133v, available http://www.bl.uk/catalogues/illuminatedmanuscripts/record.asp?MSID=764; d) source: J. Juul, Half-Real, p. 166 (modified by the author)

GAMES/EVERYDAY LIFE

Concepts of VR

The Spatial Apriori of Virtual Reality

Stephan Günzel

This contribution challenges the idea of space as an a priori form of intuition by foregrounding the limitations of Virtual Reality (VR) in HMDs, such as Meta's Quest, through the concept of a pre-given structure of spatial perception. This enables the original Kantian concept to be questioned by what Foucault termed the ›empirico-transcendental doublet‹, which was initially used to describe a historical configuration of knowledge. VR can then be redefined as a medium with a particular topology, exemplified by the Guardian that defines the boundaries of the playable space in a transcendental way while remaining empirical.

1. Virtual Reality as Stack

Before turning to the Guarding as a reflection of the empirical-transcendental double, it is important to briefly remind oneself of the origin of VR, which in fact is not only a present-day technology, but the sum of different discoveries and inventions that were made at least within the last six hundred years. It piles on to some sort of technological stack – to borrow a term from Benjamin Bratton used for the layering of the physical and the virtual – [1] that reaches back to Renaissance art and the discovery of ways to construct perspectival images. Besides some forerunners approaching linear perspective in an empirical way (like Giotto placing humans and heavenly creatures next to each other in a way that they all can be considered to share the same space),[2] towards the 16th century, namely by Viator in 1505 with the so called ›artificial perspective‹ the algorithm was discovered how to construct a coherent space prior to it being

1 Benjamin Bratton (2015): The Stack. On Software and Sovereignty, Cambridge/London: MIT Press.

2 Margaret Wertheim (1999): The Pearly Gates of Cyberspace. A History of Space from Dante to the Internet, New York/London: Norton.

filled with objects.[3] – This was done with the help of the distant point, located on the same horizon as the vanishing point, by which the transversal lines can be drawn that intersect with the lines connecting the base of the image with the vanishing point. Those intersections then allowed for the construction of the parallel lines defining the diminishing scale of the space and resulted in the typical check pattern of this era.

The next level in the stack of nowadays VR is the 360 degree-image of the panorama, which, in its modern form, was invented and named by the artist Robert Barker in the late 18th century. Even though one can find forerunners in the ancient world and also outside Europe, it can be considered to be the starting point of immersive imagery,[4] that aimed at hiding the frame, so spectators cannot tell anymore the difference between the picture plane and its outside. In the circular way those images were installed around a platform for the viewers, a ceiling above and a false terrain (*faux terrain*) around the platform prevented the limits of the picture to be visible.

A further level in the stack is not coming from the arts, but from 19th century physiology and the discovery of the stereoscopic effect: Charles Wheatstone in 1838 – the same year in which Daguerre created his first photograph – presented pairs of drawings in the style of wireframe graphics that were meant to either be viewed with crossed eyes or with a mirror apparatus that let each of the eyes of a viewer see only one picture, resulting in the synthesis of a spatial image.[5] For Wheatstone, it was the proof that vision ›receives‹ spatial information not only from proportion (smaller objects are seen as ›further away‹ than larger objects of the same kind) and occlusion (objects in the front block objects ›behind‹ them), but also from the ›layering‹ of planes. Therefore, his drawings were only outlining objects and not using surface structures or colours. In the long run, this led to the understanding of perception as an active synthesis and not a mere passive process.[6]

---

3 William M. Ivins (1973): On the Rationalization of Sight. With an Examination of Three Renaissance Texts on Perspective, New York: Da Capo Press.

4 Oliver Grau (2003 [2001]): Virtual Art. From Illusion to Immersion, Cambridge/London: MIT Press.

5 Charles Wheatstone (1838): »Contributions to the Physiology of Vision. Part the First: On Some Remarkable, and Hitherto Unobserved, Phenomena of Binocular Vision«, in: Philosophical Transactions of the Royal Society of London 128, pp. 371–394.

6 Jonathan Crary (1990): Techniques of the Observer. On Vision and Modernity in the Nineteenth Century, Cambridge/London: MIT Press.

The next step towards VR that added to the stack of the before mentioned levels, is the use of the photographic camera for recording movement on film. Just like the stereoscopic effect, which synthesises two perspectives in space, movies allow the spectator to synthesise different frames (or pictures) to become one ›movement image‹. This is most obvious in early approaches to the new medium, that did not edit the film, but rather presented a continuous shot: The most famous example is *The Arrival of a Train* from 1896 by the Lumières; but for VR even more important is the genre of the so-called ›ghost rides‹ where the camera was mounted at the front of the train and filmed the movement through space from a first-person point of view.[7]

Several systems in the 20th century then tried to fuse the existing methods in approaching VR. One is Morton Heilig's *Sensorama* from 1962, that combined (amongst other elements) ghost rides with the panoramic element of hiding the picture's frame (by the use of a ›canopy‹ that people had to stick their heads into – later on he planned also to build a head-mounted display),[8] another one is Ivan Sutherland's *Sword of Damocles* from 1968, which combined the Panorama's 360-view with stereoscopic wireframe graphics in the style of Wheatstone.[9] The latter also used another technology, and that is computer graphics. Those can be considered the last level in the stack of VR, and also something that made use of the first level in the stack: the perspectival algorithm, as it can be found in 3D-video games, most outstanding in the arcade cabinet *Battlezone* from 1980,[10] that in addition also used a view-blocker to exclude the surroundings. On addition to that, computer games bring to the stack the fundamental interactivity of the simulation image – or its possibility of the image to be manipulated by the viewers,[11] hence users.

7 Tom Gunning (1983): »An Unseen Energy Swallows Space. The Space in Early Film and Its Relation to American Avant-Garde Film«, in: John L. Fell (ed.), Film before Griffith, Berkeley/Los Angeles/London: University of California Press, pp. 355–366.

8 Morton Heilig (2001 [1955]): »The Cinema of the Future«, in: Randall Packer/Ken Jordan (eds.), multimedia. From Wagner to Virtual Reality, New York/London: Norton, pp. 239–251.

9 Ivan Sutherland (2001 [1965]): »The Ultimate Display«, in: Packer/Jordan, multimedia, pp. 252–256.

10 Steven Poole (2000): »Solid Geometry«, in: Id., Trigger Happy. The Inner Life of Video Games, London: Fourth Estate, pp. 125–148.

11 Lambert Wiesing (2010 [2001]): »Virtual Reality. The Assimilation of the Image to the Imagination«, in: Id., Artificial Presence. Philosophical Studies in Image Theory, Stanford: Stanford University Press, pp. 87–101.

2. Archaeology of the Apriori of Space

According to the critical philosophy of Immanuel Kant space is an a priori form of intuition, which means that any object that can be perceived is being seen (or heard, touched, etc.) ›in‹ space or spatially ›outside‹ of us.[12] And this is insofar ›a priori‹ as it is ›prior to‹ a particular perception. This claim is made to counter the idea of contemporary philosophers of the 18th century (like David Hume), who argued, that people experience space only from having the perception of objects, from which an idea of space is then derived. On the contrary Kant believes that space is not an effect of such an engagement with reality, but a way our mind (German: ›Gemüth‹) structures any possible perception. He describes this also as ›transcendental‹, which translates as ›the condition of the possibility of...‹. It is not immanent to perception (as the empiricists claim) and it is not transcendent (given by a superior entity), but it is just the way our perception is organized. To a certain extent this means it is ›innate‹, but more important, it is something that we cannot look at from ›outside‹ itself, since anything we observe is already ›in space‹.

The same goes for time, only that Kant calls time the ›internal‹ form of intuition. This causes problems for the following centuries in conceptualizing space and time, especially that time is prioritised over space as being ›closer‹ to the very nature of the subject (or the mind), and, furthermore, that space or objects in space is, respectively are to be considered (a) static entity/ies. This view will be overcome in the 20th century by Henri Lefebvre, who considered time as process to be inherent to space.[13] – Nevertheless, for the context of VR one can reduce the focus to space as the outer sense of perception in the context of Kant's ›Transcendental Aesthetic‹, which is the title of the opening section of the main body of his seminal publication *Critique of Pure Reason* from 1781, whereby ›aesthetic‹ has to be understood in the literal sense as regarding the *aesthesis* (which is the Greek term for ›perception‹) and not in narrow sense of experiencing artworks or judgements about particular sensations.

An important critique of the *Critique of Pure Reason* has been published already in the 18th century by a former student of Kant, Johann Gottfried Herder, who in a close reading of that opus magnum highlights all the flaws and errors

12 Immanuel Kant (1998 [1781/87]): Critique of Pure Reason, Cambridge: Cambridge University Press.

13 Henri Lefebvre (1991 [1974]): The Production of Space, Maiden/Oxford/Carlton: Blackwell.

of transcendental philosophy, foregrounding that despite the claim that one cannot go beyond the conditions of possibilities, in performing transcendental philosophy is exactly doing that.[14] And this is primarily done by using language (a topic that was central to Herder's philosophy) – one nowadays could generalize this also to the extent of the medium that is used to present transcendental philosophy. One example is Kant's separation between the internal and the external sense, implying that the inner is not spatial – and thus, ›outside‹ space. Obviously, this runs into a paradox caused by the necessity to use language – in particular prepositions (like ›in‹ and ›out‹) – to describe the setting of the mind.

Not many readers took notice of Herder's book, also because it was supposed that it was basically him taking revenge on Kant for a devastating review he published on Herder's *Ideas for the Philosophy of the History of Mankind* in which the author suggested that according to the place on earth (the culture in which you grow up) your understanding of the world might be a different one, simple due to practical reasons of engaging with for example nature differently.[15] Already in earlier research about the question why the Bible describes the creation taking place in six days plus one day of rest, Herder argued for the simple fact that this mirrors the lifeworld of an ancient culture. The method he applied in this early investigation (that only got published posthumously) Herder named ›archaeology‹ – and the title of the book should have been *Archaeology of the Orient*.[16]

Without mentioning and maybe in fact without knowing about Herder's manuscripts, in 1969 the French philosopher Michel Foucault published a book called *The Archaeology of Knowledge* in which the central chapter presents the idea of the ›historical a priori‹.[17] With this Foucault tried to describe the method he had been applying in his previous investigations, namely in his ground-breaking book *Les mots et les choses* from 1966, published in English (and other languages) under the title *The Order of Things*, which subtitle was (here all translations follow the French original) *An Archaeology of the Human Sciences*. This is the

14 Johann Gottfried Herder (2022 [1799]): Metakritik zur Kritik der reinen Vernunft, Hamburg: Meiner.

15 Id. (2024 [1784-91]): Ideas for the Philosophy of the History of Mankind, Princeton/Oxford: Princeton University Press.

16 Id. (1883): »Fragmente zu einer ›Archäologie des Morgenlandes‹ (1769)«, in: Id., Sämtliche Werke. Vol. 6. Ed. by Bernhard Suphan, Berlin: Weidmann, pp. 1–129.

17 Michel Foucault (1972 [1969]): The Archaeology of Knowledge and The Discourse on Language, New York: Pantheon Books.

first time the notion of the »historical *a priori*«[18] is used, creating – again – a paradox for transcendental philosophy, only that this time it is addressed directly and not occurring by the constellation described by Kant (and criticised by Herder).

It is very likely that Foucault was inspired by a late text by the German phenomenologist Edmund Husserl that was recognised by a wider audience in 1954 as a supplement to his monograph on *The Crisis of European Sciences*,[19] compiled from several talks Husserl gave in the mid and late 1930ies, after he was banned from teaching by the Nazi-regime. Despite this context the ›crisis‹ Husserl is talking about is not primarily a political one, but an even ›deeper‹ one that neglects the observation that the sciences are rooted in the human ›lifeworld‹, just like any other form of knowledge that once was practical before it turned ›theoretical‹, and whose most impactful results are the ›positivistic‹ sciences of the 19th century (which in turn became also the basis for political constellations like racism in the first place, based in ›scientific‹ anthropology and ethnology [example by me, not by Husserl]).

The appendix to the crisis-text was written in 1936 and first published in 1939 in a French philosophy journal by Husserl's assistant Eugen Fink as *Inquiry Concerning the Origin of Geometry: A Problem of Intentional History*.[20] This text then was translated by no one less than Jacques Derrida in 1962 and published with an introductory comment.[21] The comment as well as Husserl's original text focus on the tension that Fink highlighted as ›intentional-historical‹, because the intentional is basically what can be described in the tradition of Kant as the formal condition of intuition a priori, and the historical, which lies outside what transcendental philosophy can describe.

The link to the topic of space – and to VR in the following – is Husserl's example by which he demonstrates that tension: geometry. Which refers to geometry as it was invented by Euclid in his book *Stocheia*, Greek for the ›elements‹. Here, the elements of measuring forms in a mathematical manner,

---

18 Id. (1989 [1966]): The Order of Things. An Archaeology of the Human Sciences, London/New York: Routledge, p. 172.

19 Edmund Husserl (1970 [1954]): The Crisis of European Sciences and Transcendental Phenomenology. An Introduction to Phenomenological Philosophy, Evanston: Northwestern University Press, pp. 353–378.

20 Id. (1939): »Die Frage nach dem Ursprung der Geometrie als intentional-historisches Problem«, in: Revue Internationale de Philosophie 1/2, pp. 203–225.

21 Jacques Derrida (1978 [1962]): Edmund Husserl's ›Origin of Geometry‹. An Introduction, Stony Brook: Hays.

providing definitions and axioms, including Pythagoras' theorem and the parallel postulate.[22] What links Euclid to the starting point in Kant is the fact that the latter compared his ›revolution of thought‹ (as he himself calls his transcendental approach) to the invention of geometry by Euclid, which seems to invent a coherent system to describe and construct geometrical objects. As it is known from so called post-Euclidian geometry (coming into being towards the end of Kant's lifetime) the postulate of ›not a second line can be drawn through a point outside a given line without intersecting with it‹ is only true for an idealised plane surface that endlessly stretches out ›in space‹ without ever bending. (On a sphere many lines can be drawn through a point – like the poles on the globe – without intersecting with another line drawn on that sphere.)

For Husserl – and in turn Derrida as a possible inspiration for Foucault – it is relevant to take the historical precondition of Euclidian geometry into consideration, which literally is the ›measuring of the earth‹. Pythagoras' theorem (the sum of the areas of two smaller squares of a rectangular triangle equals the large square), for example, was discovered when ›empirical‹ reasons forced people in Egypt after the river Nile flooded the fields to measure the size of the original acres to distribute the land amongst the farmers. What nowadays appears to be a priori nests in the (historical) lifeworld.

It comes as no surprise that in the context of deconstruction, Derrida considers the reference to history as such being just another metaphysical claim, replacing the formal a priori; but what is important here is that Foucault develops the idea of the tension between the ›intentional‹ (or formal) and the ›historical‹ (or a priori) into what in *The Order of Things* he names the »empirico-transcendental doublet«.[23] Here, obviously the historical is substituted by the empirical, which causes some confusion, because to Kant the empirical is what occurs within the transcendental, so in the literal sense, what can be ›experienced‹ (in space – and time).

The reason for Foucault to do this on the one hand might be that the term of *historia* originally means ›exploration‹, which does take place within the sensual world, but more important is, on the other hand that Foucault tries to specify a certain configuration of knowledge, namely the one in which transcendental philosophy itself occurred. And the method by which this is described is the above-mentioned archaeology, which means looking for the ›ori-

22 Thomas Heath (ed.) (1926): The Thirteen Books of Euclid's Elements, Cambridge: Cambridge University Press.

23 M. Foucault: Order of Things, p. 347.

gins‹ (Greek: *arché*), in this case the origins of certain structures of thinking. Something Foucault also calls ›episteme‹, i.e. a form(ation) of knowledge.

And those origins cannot be found within philosophy, but outside of it in the way knowledge in general was organized in a given historical layer; and the episteme in which Kant's philosophy itself was possible can be described as an empirical-transcendental duplication, because what is considered to be the a priori structure of the mind by Kant was addressed by the (empirical) sciences around him as the basis of defining human beings (which then led to the invention of the Human Sciences). The three main examples of Foucault are: economy, biology and linguistics. All of them turn away from a pure description of objects that are relevant to them to the precondition of what allows them, to be such objects. Instead of calculating goods in economy, drawing taxonomies of plants and animals, or asking for what object a particular word represents, in the shift from the 17th to the 18th century, those sciences started referring to categories like ›labour‹, ›life‹, and ›language‹. Thus, instead of asking for how different languages represent objects by different words, linguists started comparing the grammar of different languages or form of articulation (no matter what the content of it is).

Whereas the former, according to Foucault, is taking place in the episteme of representation, the latter is taking place in the episteme of the empirico-transcendental doublet, and the transcendental philosophy of Kant, then, is just mimicking this structure by declaring the (empirical) properties of the mind to be transcendental. To ›illustrate‹ this shift, Foucault in the first chapter of *The Order of Things*, describes the painting *Las Meninas* from 1656 by the Spanish artist Diego Velázquez, which shows the painter himself, supposedly painting a picture of which only its backside is shown. While the canvas is turned away from the spectator, one can start coming up with some hypotheses; what actually the ›object‹ of the image in creation is (is it the girls, named in the title of the painting, standing next to Velazquez, is it King and Queen, reflected in the mirror in the back – that in fact turns out to be what could be seen on the canvas – or is it us the spectators, who Velazquez stares at when one is positioned in front of the artwork). There is no final answer to the question, but more importantly a confusion about what is represented, so representation itself is in doubt. The dawn of a new age, for Foucault, is symbolized by a person leaving the room through a door at the back of the room, entering the new episteme (of the empirico-transcendental doublet), which gives rise to the human sciences – and consequently transcendental philosophy.

From looking at the technological stack of VR one can already get an idea of how that links to Kant's claim of an a priori of space as the external form of intuition and Foucault's description of the particular episteme as an empirico-transcendental doublet. Viator's algorithm of artificial *perspective* describes the Euclidian paradigm in which Kant not only defines space as a priori to perception, but also the very comparison of transcendental philosophy itself (as the shift from empirical to formal). The very method that the *panorama* and subsequent immersive media use is hiding the frame of the picture to make the image fuse with the environment; this equals the aim of transcendental descriptions to fence itself off from any empirical content, which nevertheless is itself the condition of possibility (according to Foucault) for transcendental philosophy to make its claim. *Stereoscopy* then can be considered to be the tipping point in which the formal structure of perception turns empirical, because the fusion of the two pictures can only take place in the human being, with the result of a depth-image. The *movement image* requires the same, in the perception of a continuous experience of motion through space, and *computer graphics* finally allow for creating any possible content for VR-media, amongst which the most relevant today are head-mounted displays,[24] like the *Oculus Quest*, now renamed *Meta Quest*.

3. Setting the Guardian

There is a notable contradiction in using VR-devices and the way the experience is promoted. Usually, one sees footage of a seamless experience, without any flaws. Besides the problem of latency or bugs in applications, before the virtual experience is possible, one has to set up the device – and more important: one has to set up the virtual space within the physical space itself. With the *Quest*, this happens in form of the so-called Guardian. Its set-up here ›takes place‹ in two steps: The first one defines the floor level of the space one stands, sits, or even lies on. It is expected to make the actual floor match with the floor of the virtual space. Since the room is scanned by the headset's cameras under regular light conditions this happens automatically, but sometimes needs to done by hand, literally by touching the floor with a controller. A mismatch would result in an almost out-of-body-experience of the Avatar.

24 Jaron Lanier (1989): »Virtual Reality«, in: Whole Earth Review, Fall, pp. 108–118.

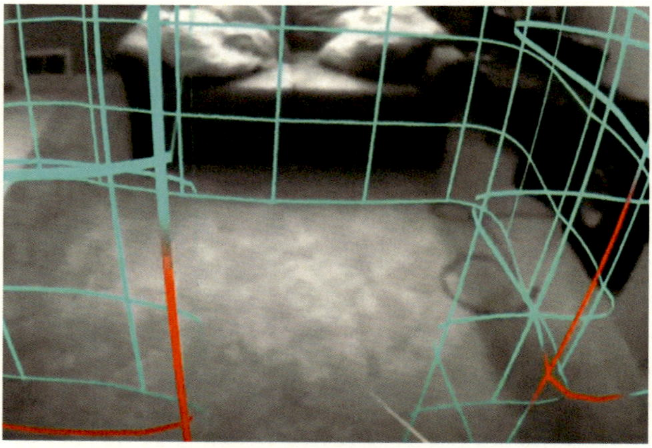

fig. 1: (Hidden) Visibility of the Guardian in the Quest

The second – and even more important – step delimits the boundary of the virtual and the physical, not according to the immanent limitations of the virtual world, but to the range within the physical world that needs to be provided to be active in the virtual. There is also the possibility to use a stationary boundary (a fixed circle around the users' position), especially for the sitting position when watching movies or playing games on virtual consoles with a game controller, without moving one's own body to move the Avatar. With the latest model, the *Quest 3*, it is also possible to have the room, or even a complete house, scanned in total, so that the boundaries of the Guardian are identical with the boundaries of the architecture and/or furniture of a space. This is what is now usually addressed as Mixed Reality (MR), which can also blend the camera feed (the pass-through of reality, now also in colour) with the virtual content or textures.

As claimed from the outset, the Guardian is not only a precondition for using VR, it also gives an insight into the topological ›nature‹ of the medium, which is essentially rooted in the epistemological structure of modernity: On the level of the dispositive, this is the technological stack, as described in chapter one; on the level of the discourse, it is the split between the actual experience and the extrapolation of the empirical as a condition of the possibility, as described in chapter two. Thus, the spatiality of VR is not only determined by the spatial geometry on the inside of the Guardian, but also by the surroundings outside the Guardian that provide the physical space for the experience. Regarding Augmented Reality (AR), this structure has already been termed aptly as the ›area‹, which »encompasses the image and its surroundings, as well as an image of the surroundings. It is the condition that makes a collision between the image and its surroundings possible.«[25] Just like in AR the image becomes foregrounded as an image when the image-object no longer fits the environment and becomes seen as an image, in VR the Guardian blends into the field of vision when the boundary between the image and the image-carrier is crossed; only that the image-carrier in this case itself is virtual (being the imaginary boundary between inside and outside).

Notably, this empirico-transcendental duplication has been addressed by art at a very early stage. This is not only the case for the above mentioned painting by Velázquez, for the crisis of representation, but it can also be found in the early days of panorama art: Caspar David Friedrich's famous *Wanderer above the Sea of Fog* from 1818 is not a rear facing portrait of a man standing on a hill overseeing a physical space, but it is an image of a visitor on a platform of a panorama painting,[26] being separated from the canvas by the fake terrain (in this case, the fog) that prevents him from crossing the boundary between the image and the medium. This does not contradict the common interpretation of it showing a situation in which the landscape is contemplated, but it gives it a medial turn, in the way that the Romantic has already considered the loss of the object it tries to reach out for. What is lost here is the actual nature that has been substituted by an image of nature. (It thus does not come as a surprise that Kant in his *Critique of Judgement* defines the experience of nature as ›sublime‹: by not being exposed to the forces of nature, one is witnessing and enjoying.)

---

25 Manuel van der Veen (2025): Augmented Reality. Für eine Kunstgeschichte der Kollision von Bild und Umgebung, Leiden: Brill/Fink, p. XXVII (transl. by author).

26 Gregor Wedekind (2021): »Philosophie der Landschaft bei David d'Angers und Caspar David Friedrich«, in: Nina Amstutz/Anne Bohnenkamp/Mareike Hennig/Gregor Wedekind (eds.), Das Bild der Natur in der Romantik. Kunst als Philosophie und Wissenschaft, Paderborn: Fink, pp. 45–76.

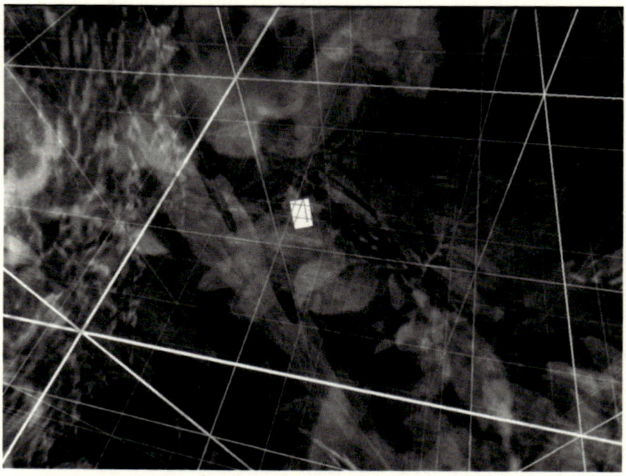

fig. 2: Transcendental grid and empirical nature in Char Davies: Osmose, 1994/95

Looking at early VR-Art one can notice a similar situation being addressed in works like the often-referenced work *Osmose* by Char Davies and her company Softimage from 1994/95 that foregrounds the grid structure of the virtual geometry being the condition of the possibility of VR to be experienced. However, in contrast to the artists' intention – as well as common interpretations based on that –[27] the attempt to regain an original experience of space or nature (with the help of a breath-controlled interface) within VR, only repeats the structure of the empirico-transcendental double and thereby is deeply ›humanistic‹. This is astonishing to a greater extend, since the presentation of the artwork refers to Gaston Bachelard,[28] with the quote »by changing space [...] we do not change place, but we change our Nature«.[29] Bachelard indeed was very critical about the geometrical understanding of space and its continued life in philosophy, but as an alternative he did not praise returning to ›nature‹,

27 Laurie McRobert (2007): Char Davies' Immersive Virtual Art and the Essence of Spatiality, Toronto: University of Toronto Press.

28 Gaston Bachelard (1994 [1958]): The Poetics of Space, Boston: Beacon.

29 Char Davies (2014): »Osmose«, in: immersence.com. Online: https://www.youtube.com/watch?v=bsT59fp8LpY (last access: 10.12.2025).

but was suggesting to think different spatialities or spatial forms (like spirals or spheres). Nevertheless, *Osmose* enables the illusion to sink below the horizon or to be immersed in the ground. This is not transgressed in most VR applications today, as it is the foundation of the Guardian.

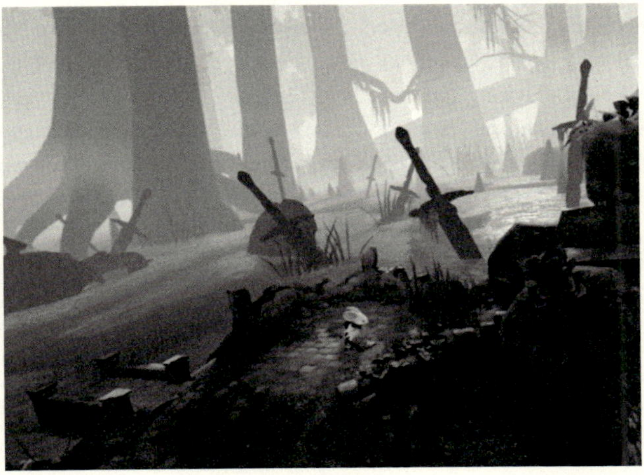

fig. 3: External perspective in: MOSS, 2018

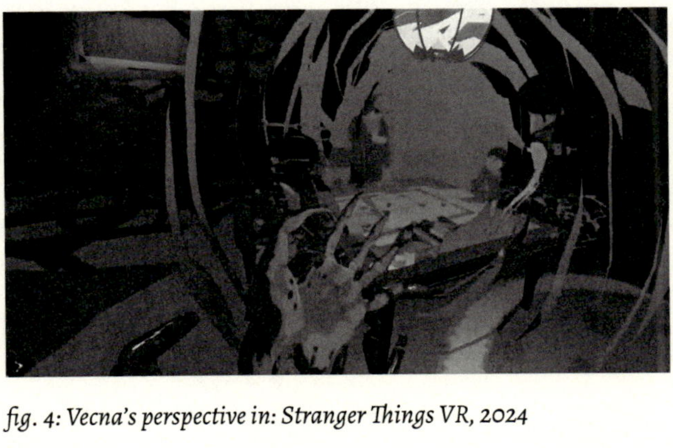

fig. 4: Vecna's perspective in: Stranger Things VR, 2024

But one cannot only detect the iteration of that very structure in artistic reflections, but also in contemporary approaches to escape the anthropocentric view in VR-games deliberately by detaching the point of action from the point of view and using supposedly non-human Avatars or lacking a player's representation at all. The first approach can be found in the Game *MOSS* (2018), in which players navigate a mouse located in an environment as well as also can manipulate that environment, too. The second can be found in *Iron Guard VR* (2021), a tower defense game, in which the players have to place gun turrets to stop the approaching forces. One of the most notable examples, however, can be found in *Stranger Things VR* (2024), in which players can take on the role of Vecna, the embodiment of evil, haunting humans from a counter-world, the ›Upsight Down‹. When reaching into the ›real world‹, the collision effect is used in a Guardian-style manner inside the game itself, whereby the outside of the players' area is the inside of the human world. Thus, the empirico-transcendental duplication is both reversed and confirmed at the same time. All these applications stick to the camera perspective as well as all the other elements of the VR-stack while trying to escape the common paradigm of VR.

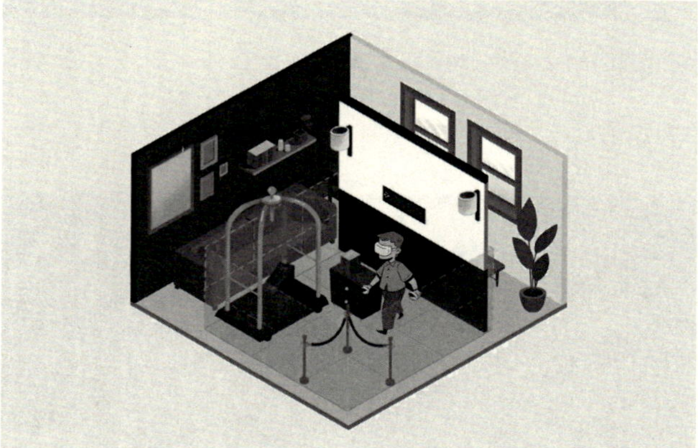

fig. 5: Infinite walking in: Hotel Infinity, 2025

Overall, the main reason for this might be the fact that the 3D-engines that are used for creating VR-experiences be default suggest the use of a camera – but beside the egocentric view also orthogonal or flat projections are offered.

Those would truly be the mode in which the above-mentioned games should be played. But designing for VR seem to not allow creators to think outside the layers of VR's stack (yet). Alternatives to think about should look into different traditions of the arts outside Europe, before the modern era, or after the heydays of photorealism.[30] Inspirations could for example come from Chinese panoramas, like the famous *Qingming*-scroll by Zhang Zeduan using orthogonal projection, or from Persian miniatures, as created by Mir Sayyid Ali, using depicting the inside and outside the same time; like in the game *Monument Valley* from 2014 for mobile devices. A very recent VR-game that started considering alternative spaces was developed by the Studio of William Chyr, who is well known for his 2019 PC and console game *Manifold Garden*, in which the game space is repeated endlessly in all directions. Inspired by the idea of transcending the limitations of the given medium, the 2025 VR game *Hotel Infinity* now requires players to walk in circles (or squares) inside the Guardian. With every turn, the interior of the Guardian can ›expand‹ or be replaced by a different bit of a virtual space that would physically have to be located outside the Guardian. Additionally, certain situations allow you to see the space you are in from outside or above at the same time. This does not yet overcome the empirico-transcendental double, but it can let users start thinking about the limits of VR – in VR.

Bibliography

Bachelard, Gaston (1994 [1958]): The Poetics of Space, Boston: Beacon.

Bratton, Benjamin (2015): The Stack. On Software and Sovereignty, Cambridge/London: MIT Press.

Crary, Jonathan (1990): Techniques of the Observer. On Vision and Modernity in the Nineteenth Century, Cambridge/London: MIT Press.

Davies, Char (2014): »Osmose«, in: immersence.com. Online: https://www.youtube.com/watch?v=bsT59fp8LpY (last access: 10.12.2025).

Derrida, Jacques (1978 [1962]): Edmund Husserl's ›Origin of Geometry‹. An Introduction, Stony Brook: Hays.

Foucault, Michel (1972 [1969]): The Archaeology of Knowledge and The Discourse on Language, New York: Pantheon Books.

30 John Willats (1997): Art and Representation. New Principles in the Analysis of Pictures, Princeton: Princeton University Press.

Foucault, Michel (1989 [1966]): The Order of Things. An Archaeology of the Human Sciences, London/New York: Routledge.

Grau, Oliver (2003 [2001]): Virtual Art. From Illusion to Immersion, Cambridge/London: MIT Press.

Gunning, Tom (1983): »An Unseen Energy Swallows Space. The Space in Early Film and Its Relation to American Avant-Garde Film«, in: John L. Fell (ed.), Film before Griffith, Berkeley/Los Angeles/London: University of California Press, pp. 355–366.

Heath, Thomas (ed.) (1926): The Thirteen Books of Euclid's Elements, Cambridge: Cambridge University Press.

Heilig, Morton (2001 [1955]): »The Cinema of the Future«, in: Randall Packer/Ken Jordan (eds.), multimedia. From Wagner to Virtual Reality, New York/London: Norton, pp. 239–251.

Herder, Johann Gottfried (1883): »Fragmente zu einer ›Archäologie des Morgenlandes‹ (1769)«, in: Id., Sämtliche Werke. Vol. 6. Ed. by Bernhard Suphan, Berlin: Weidmann, pp. 1–129.

Herder, Johann Gottfried (2022 [1799]): Metakritik zur Kritik der reinen Vernunft, Hamburg: Meiner.

Herder, Johann Gottfried (2024 [1784–91]): Ideas for the Philosophy of the History of Mankind, Princeton/Oxford: Princeton University Press.

Husserl, Edmund (1939): »Die Frage nach dem Ursprung der Geometrie als intentional-historisches Problem«, in: Revue Internationale de Philosophie 1/2, pp. 203–225.

Husserl, Edmund (1970 [1954]): The Crisis of European Sciences and Transcendental Phenomenology. An Introduction to Phenomenological Philosophy, Evanston: Northwestern University Press.

Ivins, William M. (1973): On the Rationalization of Sight. With an Examination of Three Renaissance Texts on Perspective, New York: Da Capo Press.

Kant, Immanuel (1998 [1781/87]): Critique of Pure Reason, Cambridge: Cambridge University Press.

McRobert, Laurie (2007): Char Davies' Immersive Virtual Art and the Essence of Spatiality, Toronto: University of Toronto Press.

Lanier, Jaron (1989): »Virtual Reality«, in: Whole Earth Review Fall, pp. 108–118.

Lefebvre, Henri (1991 [1974]): The Production of Space, Maiden/Oxford/Carlton: Blackwell.

Poole, Steven (2000): »Solid Geometry«, in: Id., Trigger Happy. The Inner Life of Video Games, London: Fourth Estate, pp. 125–148.

Sutherland, Ivan (2001 [1965]): »The Ultimate Display«, in: Randall Packer/Ken Jordan (eds.), multimedia. From Wagner to Virtual Reality, New York/London: Norton, pp. 252–256.

van der Veen, Manuel (2025): Augmented Reality. Für eine Kunstgeschichte der Kollision von Bild und Umgebung, Leiden: Brill/Fink.

Wedekind, Gregor (2021): »Philosophie der Landschaft bei David d'Angers und Caspar David Friedrich«, in: Nina Amstutz/Anne Bohnenkamp/Mareike Hennig/Id. (eds.), Das Bild der Natur in der Romantik. Kunst als Philosophie und Wissenschaft, Paderborn: Fink, pp. 45–76.

Wertheim, Margaret (1999): The Pearly Gates of Cyberspace. A History of Space from Dante to the Internet, New York/London: Norton.

Wheatstone, Charles (1838): »Contributions to the Physiology of Vision. Part the First: On Some Remarkable, and Hitherto Unobserved, Phenomena of Binocular Vision«, in: Philosophical Transactions of the Royal Society of London, 128, pp. 371–394.

Wiesing, Lambert (2010 [2001]): »Virtual Reality. The Assimilation of the Image to the Imagination«, in: Id., Artificial Presence. Philosophical Studies in Image Theory, Stanford: Stanford University Press, pp. 87–101.

Willats, John (1997): Art and Representation. New Principles in the Analysis of Pictures, Princeton: Princeton University Press.

Image Credits

fig. 1: Source: https://www.youtube.com/watch?v=jxxqi7AdoeQ

fig. 2: Source: https://www.immersence.com/osmose/

fig. 3: Source: https://www.pcgames.de/Moss-Spiel-23811/Tests/review-Magisches-VR-Maerchen-1251520/

fig. 4: Source: https://www.meta.com/de-de/experiences/stranger-things-vr/7803243026376133/

fig. 5: Source: https://hotelinfinityvr.com/

World[build]ing Online Exhibitions, or Trying to Resist Zombie Curating

Annet Dekker

Introduction

In the context of contemporary art, online exhibition practices are increasingly challenging conventional notions of space, time, and curatorial roles. The artist duo N*ew Scenario (Paul Barsch and Tilman Hornig)* are known for their innovative online exhibitions such as *Body Holes* (2016) and *H O P E* (2017). Using the web not merely as a distribution platform but as a site of production, their pioneering formats defy the spatial and institutional constraints of traditional gallery systems, and their exhibitions operate at the intersection of curatorial experimentation and narrative architecture. However, the conceptual frameworks available to analyse such practices often lag behind technical and aesthetic innovations. Specifically, there remains a need to distinguish between the static models of digital *world-building* and the more fluid, networked approaches of *worlding* that account for relationality, (non)human entanglement, and various forms of interactivity to embrace new ways of creating worlds. In this essay, I examine how these approaches to *world[build]ing* are manifested by focusing on space, time, and curating, and illustrated through imaginary scenarios involving an observer, a player, and the characters within the exhibition, to explore how the artist duo New Scenario constructs and builds, or *worlds*, their online exhibitions.

1. World-building and worlding

While art institutions are adapting to the digital realm, New Scenario, the German artists and curatorial duo Paul Barsch and Tilman Hornig, were quick to respond to their efforts. Profiling New Scenario in 2015 as »a dynamic platform for conceptual, time based and performative exhibition formats. it (sic) happens outside the realm of the white cube and is meant to function as an extension to create new contextual meaning«,[1] they emphasize the importance of context and critique the virtual white cube that had become a standard: a skeuomorphic representation of the physical space in which the artwork and its documentation were devalued to a low-quality image. From the beginning, their projects rejected standard exhibition protocols, in which spaces should be neutral and invisible.[2] The artworks are not presented on walls or pedestals but are embedded within hard-to-reach places, from a limousine, a dinosaur park, the contaminated zone in Chernobyl, to human orifices. Meticulously staged, the exhibitions are not opened to the public but documented for display on the web as a slide show, a sound recording, a video, or a VR environment. For New Scenario, physical exhibitions only exist to be documented, which becomes the only way the art is experienced:

> »if the documented scene and the location in which the exhibition takes place are strong enough and if the artworks and their surroundings can interact and communicate with each other, they are able to transcend the documentation and turn it into an experience that is able to replace a spatial experience.«[3]

The sensual experience of the physical space is replaced by the conceptual and narrative qualities of the documentation, providing an additional imaginative space, where anything can happen. The carefully designed, self-contained new scenarios of the exhibitions can also be seen as art projects, or ›performances‹ as they describe them.[4] Indeed, it's hard to see the exhibitions merely as attempts to present artworks, as the documentation and the translation to the

---

1 New Scenario: »Info«, in: www.newscenario.net. Online: http://www.newscenario.net/info.html (last access: 15.08.2025).

2 Carol Duncan (1995): Civilizing Rituals. Inside Public Art Museums, London: Routledge.

3 New Scenario (2021): »New Scenario (Paul Barsch & Tilman Hornig), 1 May 2019«, in: Annet Dekker (ed.), Curating Digital Art. From Presenting and Collecting Digital Art to Networked Co-curation, Amsterdam: Valiz, pp. 202–211, here: p. 207.

4 Ibid.

web create a new spatialisation, narrative and experience.[5] In this sense, New Scenario does not just curate art, it builds new worlds.

At first sight, the ways they create and present the exhibitions seem closely related to the notion of world-building. World-building is often described as a narrative practice, closely associated with speculative fiction and its various sub-genres such as science fiction, climate fiction (cli-fi), horror, fantasy and solarpunk. In other words, genres in which internally consistent, self-contained systems, or immersive environments are constructed. Typically, world-building happens through a top-down methodology in which a character exerts control over the ontological and epistemological parameters of the fictional setting.[6] Hence, world-building refers to the invention of imaginary places and societies as worlds that are built with a specific logic, structure and history. As mentioned, each of New Scenario's exhibitions is set in a specific space: a limousine in *Crash*, a dinosaur park in *Jurassic Park*, human bodies in *Body Holes*, etcetera. Rather than mere backdrops, the spaces are closely connected to the main concept. As central characters or speculative frameworks, the spaces have their own internal logic, affect, tone, and politics, which are amplified through the artworks – and vice versa. In fiction, world-building involves the creation of entirely new settings, cultures, and systems of meaning. In New Scenario's practice, it becomes a curatorial method, a design of contexts that shape not just how art is seen, but how it is experienced, inhabited and understood.

While effective in producing coherent narrative spaces, world-building often exhibits limited adaptability and these worlds are not inherently designed to accommodate emergent change, relational storylines, or collaborative intervention. Such an approach would be closer to the notion of worlding as explored in the work of Gayatri Chakravorty Spivak (1985), Anna Tsing (2010) and Donna Haraway (2016). Here, worlding can be understood as a practice of situated, relational becoming and as an ontological commitment to co-creation that offers a more fluid and process-oriented approach to imagining and constructing worlds. Next to world-building, worlding focuses on how

5 Annet Dekker (2024): »Curating Online: Art in Space and Time«, ISEA2024 Proceedings, Brisbane, online: https://isea-archives.siggraph.org/wp-content/uploads/2025/01/2024_Dekker_Curating_Online_Art_in_Space_and_Time.pdf (last access: 15.08.2025).

6 Mark J.P. Wolf (2012): Building Imaginary Worlds. The Theory and History of Subcreation, London/New York: Routledge; Henry Jenkins (2006): Convergence Culture: Where old and new media collide, New York: New York University Press.

relations between entities and things become manifest and are continuous. This reconceptualisation shifts the emphasis from world-as-object to world-as-process. It is not oriented toward the articulation of a singular, unified world but rather toward the generation of a multiplicity of overlapping and nested realities, each capable of iteration, divergence, and mutation. Worlding involves an assemblage of human and nonhuman agencies, including artificial intelligence, algorithms, and networked technologies, wherein agency becomes distributed and indeterminate. The antagonistic approach of New Scenario seems to use worlding not to impose or shape a reality to change a worldview (as Spivak would see it); rather they want to use it to disrupt conventions. Creating unstable, speculative, and strange worlds, they reject the idea of an objective or universal exhibition format to make space for fiction, affect, and alternative ontologies, thereby echoing the worlding of Tsing and Haraway. In their framings, worlding could then be seen as a socio-political act; neither neutral or purely imaginative, their aim is to present a method through which power is distributed, narratives are dismantled and alternative modes of doing or being are built. Similarly, New Scenario is not interested in a singular, unified curatorial method but wants to generate a multiplicity of overlapping and nested realities, each capable of iteration, divergence, and mutation.[7]

In line with the distinction between world-building and worlding, curator and researcher Marialaura Ghidini has pointed out the difference between »curating online« and »curating on the web«. The former relates to and mimics the practice that derives from displaying museum and gallery collections, while she describes »curating on the web« as a site-specific approach that facilitates new ways of producing and displaying digital art.[8] In other words, »curating on the web« treats the web as a medium. For Ghidini this means that it needs to reflect on the ecology of the adopted technology, in which websites are not seen as »static and self-contained objects, but rather as ecosystems that are inhabited and shaped by third parties through various interactions between

7 New Scenario expresses their views on the particularities of their exhibition methods in several interviews. New Scenario: »Press«, in: www.newscenario.net. Online: http://www.newscenario.net/info.html selection press (last access: 15.08.2025).

8 Marialaura Ghidini (2019): »Curating on the Web: the Evolution of Platforms as Spaces for Producing and Disseminating Web-based Art«, in: Arts 8/3, n.p., online: https://www.mdpi.com/2076-0752/8/3/78 (last access: 15.08.2025).

the object (the website) and its larger context«.[9] Expanding on this, curatorial researcher Gaia Tedone noticed how online curation is performed by humans and algorithms with outcomes that are emergent and contingent.[10] In these cases, curating no longer exists as the solitary, authoritative act of a curator selecting and displaying objects, but rather as a distributed, socio-technical process involving both humans (curators, artists, and users) and machines.[11] This notion, termed as »networked co-curating«, closely aligns with worlding, moving from presenting finished artworks to staging processes where visibility, circulation, and interpretation are actively co-produced between curators, artists, users and the technical environment. In this scenario, technical advances such as real-time rendering, or generative and machine-learning systems have increasingly enabled responsive curatorial modes by facilitating collaborative storytelling environments and speculative design frameworks that emphasise an open-ended narrative. In what follows, by analysing two key exhibitions by New Scenario, and focusing on the use of space and time, I will reflect on the notions of world-building and worlding to assess whether these concepts could contribute to a better understanding of how online exhibitions create worlds, as well as alternative curatorial modes.

2. *H O P E*

H O P E explores the potential of immersive technologies and pushes the boundaries of what an exhibition can mean, using online platforms not just to display, but to world. Through deliberate world-building, it invites users to navigate complex emotions, social structures, and speculative futures. This is not an exhibition space, but a proposition, a glimpse into what could emerge when everything is equally infested, or equally alive with potentiality.

9 Ibid.

10 Gaia Tedone (2017): »Co-curating with Cassini: from the Abyss of Commodification to the Exploration of Space Curation«, in: Marialaura Ghidini/Rebekah Modrak (eds.), #exstrange: a Curatorial Intervention on eBay, Ann Arbour: Maize Books, pp. 180–185.

11 Annet Dekker/Gaia Tedone(2019): »Networked Co-curation: an Exploration of the Sociotechnical Specificities of Online Curation«, in: Arts 8/3, n.p., online: https://www.mdpi.com/2076-0752/8/3/86 (last access: 15.08.2025).

Part 1. Space

What are the potentials of online exhibitions spaces when leveraged for speculative world-building, moving beyond their traditional role as mere display platforms? I first encountered New Scenario's exhibitions on their website around 2016. Their project *Body Holes. If the body were a museum there would be seven galleries* was part of the Berlin Biennale 2016. *Body Holes* presents miniature artworks in the seven openings of the human body: from the ears, nose and mouth to the genitalia.

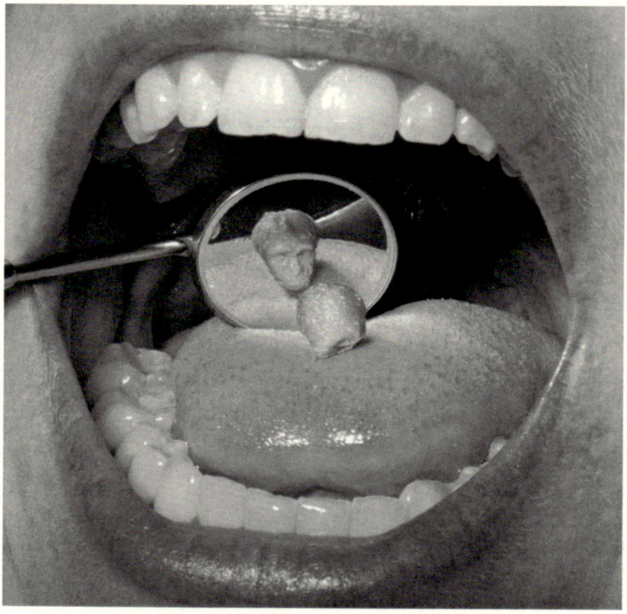

fig. 1: New Scenario: Body Holes – Paul Barsch, Homo Heidelbergensis (mirror), 3D print, 2015

By abstracting the body's orifices to this extent, the project questions the stigmatisation of the human body and addresses, in particular, the (ab)use of the body as a social and political space in the arts. However, the project is not only about breaking existing taboos surrounding the body, it primarily aims to shift the thinking and scale of an exhibition space from external to internal by asking if *every imaginable ›space‹ could be a space for art?* This absurdist way of presenting art, which evokes at the same time a sense of disgust, horror, beauty and surprise, could be seen as mere attention-seeking and an example of »extreme sports« curating.[12] Yet, by interweaving narrative texts on socio-political issues of gender, privacy or surveillance, their focus is on the broader context of how images, art and politics are discussed in the artworld.

For the exhibition *Body Holes*, New Scenario asked the artists to present existing artworks or create new ones for intimate and metaphorically charged bodily orifices. Most of the artists responded by engaging directly with the assigned ›holes‹, producing works that they either incorporated within or juxtaposed against their symbolic sites, thus inviting new interpretations. By documenting the artworks in macro photography, which provided a clinical, detailed, and objective perspective, the project neutralised the affective charge of the orifices and, in doing so, de-stigmatised perceptions of the body. In other words, it is through New Scenario's approach to documentation that even unorthodox spaces are freed from their constrained cultural, political, and sexual associations, positioning them instead as neutral and even aesthetic environments. Yet transforming the body into a seemingly sanitised exhibition space is not without ambiguity. In presenting these bodily interiors in a sanitised, detached manner (in some cases enhancing this by using a dental mirror to provide a more spatial view of the object in the mouth), New Scenario arguably follows, and perhaps even reinforces, the logic of purification associated with the modernist white cube. As art critic Brian O'Doherty observed, the art gallery is »unshadowed, white, clean, artificial – the space is devoted to the technology of esthetics«.[13] Indeed, after the initial shock and becoming familiar with their spatial context, when looking at the images the orifices recede into the background, turning into visual surfaces, landscapes, pedestals, or aesthetic wallpaper, against which the artworks emerge, much like a firehose in a modern museum can accidentally turn into an aesthetic conundrum.[14] However, moving back and forth between setting and art, the artwork starts to merge with its surroundings, forming a new image and an additional layer of meaning in which anus and art enhance each other.

---

12 A. Dekker: Curating Online, n.p.

13 Brian O'Doherty (1976): Inside the White Cube. The Ideology of the Gallery Space, Santa Monica, CA: The Lapis Press, p. 15.

14 Ibid.

Perhaps this is why New Scenario tried to create an even more preposterous backdrop for their art in their next exhibition *H O P E* in 2017. The intentionally optimistic title obscures the horror of its *mise-en-scene*, if only for a minute while the VR project loads on either the computer or the cardboard VR viewer.

HOPE

VR game space. Loading. Half of the screen slowly circles around, showing large glass windows facing a distant ceiling, interspersed with theatre spotlights mounted on scaffolding. A circle in the centre of the screen displays the names of artists – red on black.

Observer

The observer cautiously examines the scene, holding the computer mouse firmly in the left hand, while the other grips a pen. The cursor moves slowly through the virtual space, shifting left, turning right. When it encounters people or objects, the index finger clicks, hoping to reveal an artwork.
The space is in state of chaos: people are scrambling for a safe space, crawling on the floor, climbing over furniture. Blood is everywhere. There is lots of noise without any sound. A lecture hall constructed like a panopticon. The observer follows a path to the right, a lecturer points to the board, drawing his distressed audience's attention to the text reading: »proposed structure for meme warfare centre«. Clicking the dot, a pop-up video appears: a lecture by Daniel Keller, The Basilisk (2017).

Player

Putting my portable into the cardbox viewer, I quickly glance around the space: what is happening, where are the main characters, where are the clues – how do I find the artwork.
It's a gruesome scene. I'm falling down from the top into anarchy, blood splatters across the floor, blood-written graffiti, a lot of evidence of past mayhem. I speed across, moving my head left to right, up and down. In the right corner, I spot the art, quickly scanning it: »memes, warfare, EMC external meme centre« – hmm, makes sense in this ordeal. My quest continues... got it! Pulsating circles is what I'm after!

Art/Artists

Here we are: crisscrossed blood splatters on the wall, stains on the carpet, papers scattered across the floor, people stumbling over fixed tables and chairs, blank stares –

then, looking straight at us, the lecturer in front of the blackboard, pointing at The Basilisk:

> »a fabled chimeric reptile, known to be king of serpents and said to have the power to kill anyone with a single glance. According to one version of the legend, the only person to successfully kill a basilisk did so by wearing a suit of mirrors which reflected its toxic gaze back at the basilisk, turning it into stone«.

Zigzag out of here…

Using the flexible software krpano, which is designed to display all manner of panoramic images on the web, *H O P E* presents a series of detailed high-resolution images that are connected through links creating various pathways through the building. The interactive 360° virtual tour is set inside the Technical University Dresden. Commissioned by Gwendolin Kremer, curator of the Altana Gallery at the Technical University in Dresden, the same city in which New Scenario graduated from the art academy a few years earlier in 2015, *H O P E*'s familiar setting of a university is crudely disrupted by a zombie apocalypse. Viewers can navigate 17 different rooms: from an auditorium, to laboratories, toilets, hallways, canteen, study spaces, library, offices, and an elevator. In each of the spaces, they can encounter an artwork: a pop-up video, a painting or drawing, or a text. However, these are sometimes hard to find amidst the horror. By scrolling on the landing page of the website, the visitor spirals down from ceiling height to zoom into the crotch of a student-zombie laying on the floor. Double-clicking zips the viewer through the space to land in front of what seems to be the face of a student who is wearing headphones and holding a text in his left hand, with a pencil in his right to make notes. He stares sternly into the camera. After the initial flying across the room, the navigation becomes easier, simply using the arrow keys and finding the pulsing dots for more information, perhaps an artwork, or an entrance into another space.

Navigating to where the first student-zombie is lying on the floor, a girl with a mobile phone is crouching beside him, either trying to make a call, send a text or, more likely, take a photo of him. However, clicking on the pulsating circle of her mobile opens the credits: slowly scrolling red text on a black background accompanied by Choir! Choir! Choir!'s 2016 cover of the song *Zombie* by The Cranberries, its chorus echoing the massacre and message in *H O P E*:

> In your head, in your head
> Zombie, zombie, zombie-ie-ie
> What's in your head, in your head?

Zombie, zombie, zombie-ie-ie-ie
Oh-oh-oh-oh, oh-oh-oh, eh-eh-oh, ya-ya[15]

Back in the lecture hall, the invasion continues, not as a singular event but as a perpetual state, slowly unfolding in an enduring zombie modus. The scene is set and the message is clear: we are spectators and participants in a world on the brink. It mirrors the familiar aesthetics of the *Living Dead* films. Yet, the choice of 360° still images, rather than film, places the viewer in a position of pseudo-agency: you are able to look around without being attacked. It is immersive but frozen: the moment of catastrophe is held in permanent suspension. The stillness, in which nothing moves and everyone is depicted in one freeze, is indeed less cinematic and closer to the painterly abyss of a Bruegelian allegory. Trapped within the confines of the precisely constructed tableau of *H O P E*, the images come to life by moving the cursor around the space, and contemporary societal anxieties emerge as they are executed by ›brushstrokes‹ of code. Most VR and interactive projects offer players multiple viewpoints to choose from, either to determine the narrative flow, alter the space by moving or destroying objects, by building structures, or affecting non-playable characters (NPCs). Similarly, an allegory involves shifts in narrative and perspective. These stimulate different interpretations of the presented themes, both within and beyond the world that is depicted. The idea that the more a player can do, the more dynamic or immersive the process of worlding becomes, may provide them with a feeling of real-time agency. However, navigating and exploring a space may also feed a pseudo-agency that moves beyond the confines of the virtual experience. *H O P E* presents a liminal world, constructed, curated, and coded, which is inhabited by social and political narratives that read like visual statements, each with their own mise-en-scene in which the characters seem strangely at ease in the world they inhabit. The back-and-forth between these multiple and often contradictory perspectives is what forms the core of worlding in *H O P E*.

15 *Zombie* (1993) by The Cranberries reflects on the decades-long conflict in Northern Ireland between nationalists (mainly self-identified as Irish or Roman Catholic) and unionists (mainly self-identified as British or Protestant). It's dedicated to two young boys, Tim Parry and Johnathan Ball, who were killed in an IRA bombing in Warrington, England, on 20 March 1993.

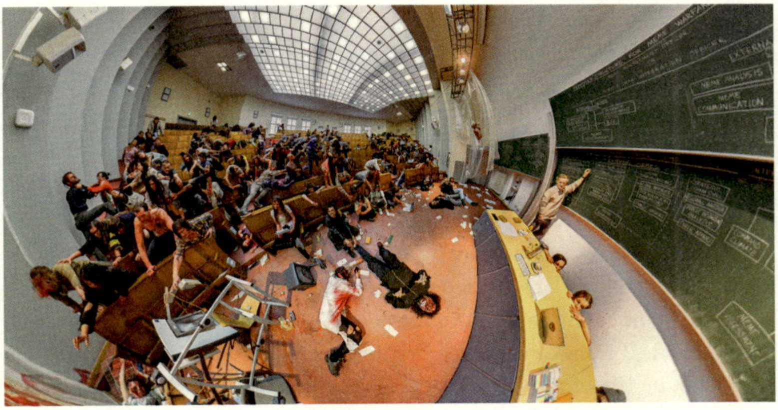

fig. 2: New Scenario: HOPE – Daniel Keller, Proposed Structure For Meme Warfare Center (MWC), 2017

In a similar way, the artworks are protagonists that carry the emotional and ideological weight of the setting. Often presented in pop-ups, the individual artworks can be viewed, read or listened to. Some artists adapted their work specifically to the setting, like Jon Rafman's video contribution *Dream Journal (Hope)*, in which he weaved together, via an incarnation of camgirl Ann Lee, a deep-web imagery of repressed libidinal fantasies, masturbating androids, hypermasculine furries, and vore communities, to create a vision of an internet addict's subconscious mind.[16] Others were selected for their resonance with the thematic architecture of the university environment, such as Jason Hirata's *Puyallup (1942 & 2002)*, a series of prints on security paper naming every dish served for one week at the concentration camp in Puyallup, Washington, where his grandparents were interned during World War II – of course, presented in the deserted canteen strewn with unfinished meals (many fried eggs), the leftovers of an abrupt departure.[17] Or, Daniel Keller's *Proposed Structure For Meme Warfare Center (MWC)*, a chalkboard diagram that starts the

16 For a more detailed description see Pierre-Alexandre Mateos/Charles Teyssou (2017): »Zombie Subjectivity: New Scenario«, in: https://www.moussemagazine.it/ (18.09.2017). Online: http://moussemagazine.it/zombie-subjectivity-new-scenario-2017/ (last access: 15.08.2025).

17 For more information, see Zoey Lubitz (2020): »Art As Negotiation: Jason Hirata Interviewed. Artwork that complexifies the notion of authorship«, in: https://bombmagaz

VR adventure. Clicking on a dot on the chalkboard opens the link to a video lecture in which Keller discusses Meme Magick, known from the alt-right, pointing out how right-populist memes are reminiscent of undead images. Finally, the soundscape of the exhibition fills the space with contradictions: familiar administrative noises, such as ringing telephones, are interspersed with car honks, sounds of the slasher's circular saw clashing with harpsichord melodies, as well as calming insect rustles alongside thunderous storms. These sonic layers mimic the absurdity of a collapsing civilisation: is it half-bureaucratic nightmare, half-baroque opera, half-ecodisaster, or the latest version of I-Doser? The sounds act as a counterpoint and a companion to the visuals, underscoring the tension between dystopian imagery and nihilistic irony. The contradiction is intentional: a death wish masked as hope, or perhaps hope is only imaginable through total collapse. Barsch notes that the artworks only reach their full narrative potential in the context of the world they inhabit.[18] In this sense, the artworks become *H O P E's* most important characters.

Within the online format, the artifice of *H O P E* is transparent. Similarly, the detailed setting, including numerous references to online behaviours, shows that New Scenario is critically aware of how disaster is staged and performed. Hence, *H O P E* doesn't simply reference disaster; rather, the deliberate aesthetic invites reflection on the medium's role in constructing and perpetuating states of emergency. Likewise, the digital space is not merely a gallery substitute, but a stage for speculative world-building. *H O P E* explores the potential of immersive technologies and pushes the boundaries of what an exhibition can mean, using online platforms not just to display, but to world. Through deliberate world-building, it invites users to navigate complex emotions, social structures, and speculative futures. This is not an exhibition space, but a proposition, a glimpse into what could emerge when everything is equally infested, or equally alive with potentiality.

---

ine.org (18.02.2020). Online: https://bombmagazine.org/articles/2020/02/18/art-as-negotiation-jason-hirata-interviewed/ (last access: 13.10.2025).

18 Matt Welch (2017): »Crisis and Infection: An Interview with New Scenario on Chaos + Positivity Ahead of Their Upcoming Project H O P E«, in: www.aqnb.com (01.05.2017). Online: http://www.aqnb.com/2017/05/01/crisis-and-infection-new-scenario-talk-chaos-positivity-ahead-of-theirupcoming-project-h-o-p-e/ (last access: 15.08.2025).

Part 2: Time

How does the potentiality of online space connect to how (digital) time is experienced? This is a philosophical, material, practical and cultural question, and the understanding of time has undergone significant transformations, particularly with the advent of new technologies and acknowledgement (in the West) of other notions of time, which have redefined how it is experienced and measured. Whereas Lewis Mumford's description of time, »The clock, not the steam-engine, is the key machine of the modern industrial age. For every phase of its development the clock is both the outstanding fact and the typical symbol of the machine: even today no other machine is so ubiquitous«,[19] is still valid, the notion of time is also cultural and socio-political, as mentioned by Meryem-Bahia Arfaoui:

> »One evening, my uncle and I sat in our courtyard in Jendouba in north-western Tunisia, talking about our homeland. In Tunisia, ›we got our independence in 1956,‹ I said. He abruptly shot me a stern look, almost as if by reflex, before retorting, ›1956... that's a French date!‹ [...] time and chronology are not politically neutral. And colonizer time is not the same as time kept by the colonized.«[20]

Computer time is also not chronological, it operates at different time scales. Even trying to program time into the computer is actually quite difficult. One has to take into account all kinds of specifics: distinctions between shifting global summer and winter times, historical anomalies, political differences within the same city (think of Israelis following a different time from Palestinians, even in the same city), and the notion of leap time.[21] Several people have proposed the detemporality of internet time, to signal how the acceleration of information resulted in a temporal compression in which time as a process

19 Lewis Mumford (1934): Technics and Civilization, London: Routledge, p. 14.

20 Meryem-Bahia Arfaoui (2021): »Time and the Colonial State«, in: https://thefunambulist.net (21.06.20221). Online: https://thefunambulist.net/magazine/article/time-and-the-colonial-state (last access: 15.08.2025).

21 Tom Scott explains the convoluted part of how while programming time: »time twists and turns like a twisty-turny thing. It's not to be trifled with!« Tom Scott: »The Problem with Time & Timezones«, in: www.youtube.com (30.12.2013). Online: https://www.youtube.com/watch?v=-5wpm-gesOY (last access: 15.08.2025).

disappeared.[22] Such observations tend to see computational space as fluid and continuous; a nice uninterrupted flow of data and exchanges. However, in most experiences this space is more volatile, less like a steady stream and more like a wild river that moves at different speeds, including downtimes, blockages, missing parts, relays and dead ends. In that sense, rather than detemporality, it is a temporal incoherence.[23]

Indeed, computer time is less a topographical or a temporal division because it corresponds to its technical specification. This is particular visible in online exhibitions, where temporality is explored through generative art, live data streams, or webcams, or where users are given control: offering downloadable ZIPs, customisable templates, or collaborative tools like Google Docs. Such formats invite participation and reinterpretation and will evolve, embracing the fluid, unpredictable rhythms of the web. By leveraging algorithms, platform features, and co-creation, such exhibitions challenge traditional notions of time, proposing alternative temporalities shaped by human and machine logic.

HOPE

INFO: krpano 1.19-pr8 (build 2016–09-30) INFO: Android 6.0.1 (Nexus 5X)–Chrome 133.0–WebGL INFO: Registered to: BTSA. CLOSE. LAYERS. HOTSPOTS.

Observer

Visitors navigating through this world lose their sense of time. This happens on the level of the space: the floorplan mimics a labyrinth of rooms presenting scenes that provide little foothold. The exhibition: a selection of artworks ranging from 1942 to the contemporary. The duration: VR contributes to time compression, we forget time.

Player

I'm moving my head, clicking the dots endlessly. I must have passed through all of the spaces. Not sure if I missed some I-Doser. The time, what is the time, I must still have a few minutes to find my way out. Wait–wait?–what is the time, how long have I been here... Where is the exit...

22 For instance, Manuel Castells (2010): The Rise of Network Society. The Information Age. Economy, Society, and Culture, Vol. 1 (second edition), Oxford: Wiley-Blackwell.

23 A. Dekker: Curating Online, n.p.

Art/Actors
Standing still. We wait. When will we be able to move, into the future, into the past. Here is only the present. Our world is the now, we are present.

Although it could be said that online exhibitions, unlike their offline counterparts, remain online (until they are no longer updated), even if only existing in a state of stasis; they are always of the present. However, from a media-archaeological perspective, technology creates a folding in time in which new forms of temporalities emerge, as explained by Sybille Krämer, in reference to Friedrich Kittler:

> »What is unique about the technological era (from the gramophone to the computer) is that these technologies allow one to store ›real time‹ – in other words, those processes that cannot be fixed by syntactical structures and are thus not irreversible, but rather contingent, chaotic, and singular – and, at the same time, to process ›real time‹ as a temporal event. *Data processing becomes the process by which temporal order becomes moveable and reversible in the very experience of space* (Kittler, 1997:130–46).«[24]

Emphasizing processing over storage shows how online exhibitions are volatile rather than stable, creating their own temporality.[25] Understanding the web in terms of the processes that shape its structure, including storage and memory, has implications for how time is experienced and conceptualised online. As stated by Wolfgang Ernst, »In technologies, there is no present at all – rather differential tempor(e)alities«.[26] According to Ernst, the symbolic ordering of

24 Sybille Krämer (2006): »The Cultural Techniques of Time Axis Manipulation. On Friedrich Kittler's Conception of Media«, in: Theory, Culture & Society 23/7-8, pp. 93–109, here: p. 96, https://doi.org/10.1177/0263276406069885.

25 This is well-known phenomenon in game play, and particularly in VR games, where the technology influences a player's notion of time, as recalled by psychologist Grayson Muller: »I stopped playing the game, and I realized that I had no idea how much time had passed. […] I was supposed to be taking turns with other people, and I was worried that I had played for too long because I couldn't even guess if it had been 10 minutes or 40 minutes«. Public Affairs (2021): »Virtual reality warps your sense of time. Psychology research demonstrates the unique ›time compression‹ effect of virtual reality«, in: https://news.ucsc.edu (12.05.2021). Online: https://news.ucsc.edu/2021/05/virtual-reality-time-compression/ (last access: 15.08.2025).

26 Wolfgang Ernst (2017): »The Delayed Present: Media-Induced Tempor(e)alities & Techno-traumatic Irritations of ›the Contemporary‹«, in: Geoff Cox/Jacob Lind (eds.), The Contemporary Condition, Berlin: Sternberg Press, p. 11.

time is replaced by overlapping and concurrent temporalities, where everything appears to happen now; more precisely, where the ›now‹ is stretched, layered, and reactivated. In this logic, there is no singular present and time becomes a recursive loop, or perhaps a zombie; undead and unchanging until disrupted, ›slayed‹, or broken. In this way, and similar to the allegoric scenario, internet time becomes a kind of eternal present, a dynamic *and* frozen moment sustained by the processes of execution.

This framework is reflected in the rhizomatic structure of *H O P E* that employs a nonlinear curatorial logic. Its spatial arrangement defies chronological narratives; instead, it offers a transversal, zigzagging experience. Similar to space, time is not a neutral backdrop. Together they become dynamic agents in the staging of experience, i.e., producing a present. This is not a singular, fixed moment of presence, but a multiplicity: coexisting moments, imagined human presences that may be familiar or alien, rooted in the past or projected into the future – whatever lies behind the next door. These moments reflect not only the temporal multiplicity of perception but also the philosophical dimension of *presence as presentation*. To present presence is to make palpable the folds in time: actual and virtual, experienced and potential. In the philosophy of becoming, memory is never static. It is multidirectional, operating through acts of recomposition. Indeed, when opening one of the doors in *H O P E* you are never sure if it leads to a new, or an already visited space. The past is not merely recalled but returns in anticipation of the future. It is enacted in both the actual and virtual present, like a zombie who is resurrected and repurposed in new contexts, while also remaining the same. As Brian Massumi explains in his book *Parables for the Virtual: Movement, Affect, Sensation (2002)*, the virtual is connected to a state of becoming, a kind of pre-actualised potential that influences and informs what is actually happening in the present (the actual). The actual present is always permeated by the virtual, which represents a potentiality: a process of becoming, or a »folded« aspect of experience that is not yet fully actualised. In VR this concept takes a unique form, because the past is not simply remembered or reconstructed as it would be in a book or a film. For instance, in *H O P E* the VR experience doesn't just exist in abstraction, rather it draws on the actual (the present world and the past) and reintegrates it into a space where it is re-enacted, re-experienced, and re-imagined. Hence, *H O P E* evokes a layered relationship between time, memory, and (pseudo-)agency in the digital space. The past (represented by the return of the zombie, sometimes in replica) is not merely a distant past or memory, but is repurposed in the present. The navigation that happens via numerous doors, endless corridors, a flickering lift, and rows of toilets (all indicators of movement), deepens this notion. These elements suggest that the past does not merely come back in recollection, but it is re-experienced in real-time. This return is not static either: the user is invited to walk through and engage with the past as if it exists in the present.

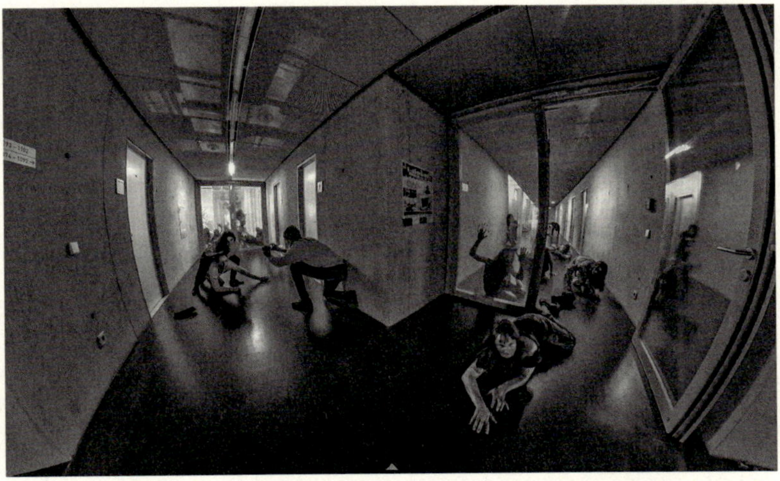

fig. 3: New Scenario: H O P E – Kareem Lotfy, I-Doser & (BRING YOUR OWN DOPE), Posters, 2017

Finally, *H O P E* as a technical system is also a medium for unfolding temporalities[27]: an enduring interface between memory, presence and the continuous becoming of time itself. As before, the 2D renderings in the VR/360° environment are not merely static, together they emphasise the notion of the actual: a convergence of the present reality, which is shaped by past decisions, events, and experiences, but also influenced by future potentials and possibilities. This effect is further enhanced by the presentation of the artworks within these spaces, which with each encounter overwrite existing meanings, inviting renewal and rewriting: the exhibition is re-presenced, producing an extended present, a compressed time window where conventional Western distinctions

27 Or, as mentioned by Ernst, new forms of tempor(e)alities, ibid.

between past, present, and future lose their linear order. Playing with these various modes of temporality, *H O P E* suggests a radical potentiality in curatorial practice, aesthetics and political engagement that is fundamental to the notion of worlding.

Part 3: Curating

Curating is everywhere. The terms curator, curating and digital curation have become all the hype: everyone is a curator and anything can be curated, from online marketing to coffee shops, children's parties, energy bars, earrings, etcetera. Along the way, more conventional fields, such as information science, chipped in, rebranding traditional terms like archiving and recordkeeping as digital curation, further blurring the lines between art, information, and commerce.[28] The number of online galleries has been increasing over the last few years, most with obscure names, such as vFairs, Omeka, Artsteps, or VOMA, together with the rise of web builders, social media platforms, and design software like SketchUp, WallApp, or 3D modelling programs like Rhino, as well as a proliferation of online exhibition tutorials: all made available to democratise the trade. Once confined to physical institutions and galleries, the curatorial methods now omnipresent enable anyone with the required skills and internet connections to present content to a global audience. These shifts have made curating an open activity, no longer exclusive to professionals in traditional institutions.

Paraphrasing Andrew Dewdney's take on photography, what happens to curating when it is everywhere, but not as we have known it: for some time the curator has been an undead, a zombie, in which the established language, thinking, meanings and values of curating stand as an obstacle to understanding its new condition.[29] The current mode of online presentation, circulation

28 A. Dekker: Curating Online, n.p.

29 In a similar way, the notion of ›zombie media‹ is used by Hertz and Parikka to address the living dead of media culture, however, referencing Bruce Sterling's concept of »dead media«, they mentioned that focusing on the forgotten or obsolete media is a useful way to »understand better the nature of media cultural development«. Garnet Hertz/Jussi Parikka (2012): »Zombie Media: Circuit Bending Media Archaeology into an Art Method«, in: Leonardo 45/5, pp. 424–430, here: p. 429.

and algorithmic interferences turn curating on its head and with it, is changing the role of curators, as well as the institutions from which they originate.[30] Like photography, curating has been freed from its conventional constraints to haunt the opaque intimacy of the web and its computational abstraction, creating new challenges to understanding curating in computational and network culture.[31] Although the question of how to regard the curator's afterlife is beyond the scope of this article, Dewdney's remark and method helps to redefine the notion of the curatorial in New Scenario's *H O P E*. Casting the afterlife of the curator as the fictional figure of the zombie curator, and positioning curating as a zombie method, is both a playful and a methodological way to discuss the limits of the institutional curator in contemporary cultural institutional practices (echoing the critique of New Scenario), as well as to understand its online equivalent.[32]

HOPE
A concept. Formulating criteria and making selections. Painting, sculptures, pedestals, lighting, distance, all design elements. Writing captions, press releases, and wall labels. Sending invitations. Documenting the scene, with and without visitors.

---

30 See, next to the earlier mentioned (n. 8, 10, 11), for instance, Joasia Krysa/Magda Tyżlik-Carver (eds.) (2025): Curating Superintelligences: Speculations on the Future of Curating, AI and Hybrid Realities, London: Open Humanities Press; Annet Dekker (ed.) (2021): Curating Digital Art. From Presenting and Collecting Digital Art to Networked Co-Curating, Amsterdam: Valiz; Magda Tyżlik-Carver (2017): »| Curator | Curating | the Curatorial | Not-Just-Art Curating: A Genealogy of Posthuman Curating«, in: Springerin, The Post-Curatorial Turn 1, online: https://www.springerin.at/en/2017/1/kuratorin-kuratieren-das-kuratorische-nicht-nur-kunst-kuratieren/ (last access: 15.08.2025); Joasia Krysa (2014): »Can Machines Curate? Notes on History of Technology Hope and Paradoxes of Curating«, in: Y. Guasque (ed.), Digital Art: Fractures, Proliferative Preservation and Affective Dimension, 5th National Symposium of the Brazilian Association of Cyberculture Researchers, MediaLan/UFG, Funape; Olga Goriunova (2013): »Light Heavy Weight Curating«, in: Annet Dekker (ed.), Speculative Scenarios: Or What Will Happen to Digital Art in the (Near) Future, Eindhoven: Baltan Laboratories, pp. 25–32.

31 Here I'm echoing Dewdney's reference to photography, and applying the same mechanism to curating. Andrew Dewdney (2021): Forget Photography, London: Goldsmiths Press, p. 3.

32 Ibid, p. 10.

Observer
I'm navigating through the different spaces, entering each of them and locating the artworks by clicking on pulsating dots. I can see how the narrative slowly unfolds. The zombies embody eternal time, trapped in their unyielding cycle – will I become part of them, or am I the barrier to their progress?

Player
The space doesn't move. Even the zombies are motionless. No action, except my own. Perhaps some of the artworks will offer more clues on how to escape from this endless misery.

Art/Audience
A zombie is not quite dead, it refuses to die – it can be a metaphor for the institution (the museum gallery that persists without a real function in the present day), the curator (whose ego is prominent, or hiding behind various facades in which the art plays a secondary role, if at all), the audience (wandering around without purpose or direction), or perhaps all of these together? What happens when considering the notion of zombie curating, or the long afterlife of the museum as institution?

The title *H O P E* signifies different things. Art critic Matt Welch recalled Barack Obama's 2008 US election campaign that used »Hope« as its slogan. In their interview, Barsch explains it as a gesture of »nihilist irony«, albeit, the title was chosen at the end of the process as an attempt to evoke a positive message to compensate for the horror that is presented in the exhibition scenes.[33] *H O P E* can also reflect the hopelessness that seems omnipresent in today's society, which as Barsch points out, in »extreme ways can manifest in a death wish hope that only total chaos and collapse can bring change and a better life«.[34] Almost ten years later the tension between hope and despair has become mainstream with a rise in books and workshops around self-help, mindfulness and meditation alongside survival courses and (governmentally advised) prepping guides to survive an apocalypse. Horror authors often argue how horror and hope are connected, and research has shown how horror gives hope and helps build resilience,[35] but what does it mean to wallow in horror or identify with the zombie? More specifically, and to stick with the topic of curating and worlding,

33 M. Welch: Crisis and Infection, n.p.

34 Ibid.

35 This notion builds on a long tradition, among others, from Friedrich Nietzsche, who wrote about creating a new tragic culture (The Birth of Tragedy, 1872), to the more re-

would it be possible within these manifestations to find new ways of curating? What would it mean to break away from Christianity's enduring legacy of compulsive hope, or in this case, ubiquitous institutional curatorial standards?

As mentioned, with *H O P E* the artists introduced an additional theme to their exhibition design, linking the exhibition spaces to a unifying concept of the zombie. While in Western culture the zombie is best known as a manifestation of the walking dead, the origin of the word is often acknowledged as coming from countries in West Africa: »West African Vodun holds that a dead human can be brought back to life by a priest with specific otherworldly power to commune with the gods. This priest, or *bokur*, then gains control of the zombified human, who has lost all free will«.[36] On the other hand, the astral zombie, one whose soul has been captured, derived from »nzambi« in Kikonko, one of the Butan languages, can in this reading also connote good luck and literally means God.[37] The enslaved people from West African countries brought both notions of the zombie to Haiti and the United States of America, where they gained popularity in literary and popular culture. However, in line with colonialist ideology, the Western notion of the zombie became best known for its portrayal of *otherness* – figures who had lost their ability to speak, lacked will or expression, and existed only to haunt the living.[38] Hence, and despite the wide range of zombie figures, genres, and myths from literature, films and television series, and social media doom scrolling, one thing they all have in common is their subdued and dominated sense of agency. Being under the spell of their conqueror, they live a sedated life.

Similarly constrained by the demands of the institute, and partly under the influence of a critical rethinking of the institutional, the 1990s saw the emergence of the independent curator detached from organisational affiliations (like museums and galleries), and able to operate transnationally.[39] Experimenting with alternative formats of display and engagement, their

cent book by Olga Goriunova/Matthew Fuller (2019): Bleak Joys, Minneapolis/London: The University of Minnesota Press.

36 June Michelle Pulliam/Anthony J. Fonseca (eds.) (2014): Encyclopedia of the Zombie: The Walking Dead in Popular Culture and Myth, Greenwood: Bloomsbury Publishing, p. 2.

37 Ibid.

38 Edward P. Comentale (2017): »Zombie Race«, in: Sarah Juliet Lauro (ed.), Zombie Theory: A Reader, Minnesota: University of Minnesota Press, pp. 189–211.

39 The independent curator also gained prominence due to the globalisation of the art world and the proliferation of biennials and large-scale exhibitions that required mo-

work highlighted power structures within the art world and questioned who gets to define artistic value and cultural narratives. Curator and researcher Paul O'Neill suggests that independent curating became a form of cultural production, creating new meanings and experiences rather than simply managing or preserving existing works. However, he also acknowledges that no curator is ever truly independent, as they often remain embedded in broader cultural, economic, and institutional frameworks.[40] In the early 2000s, the widespread and mainstream adaptation of the notion of curator also signaled how curatorial practices have become commodified. While the notion of the curatorial became accepted throughout various practices, it often reproduced the very hierarchies and market dependencies that independent curating initially sought to resist. Similar to the zombie legacy in photography that is haunting the understanding of contemporary images,[41] ›zombie curating‹ creates a state of curating where its established language, thinking, meanings, and values have become an obstacle to understanding its contemporary condition, functioning instead as an ›undead‹, or ›zombie‹, entity. Being part and parcel of the long afterlife of the museum as institution and the challenges faced by curatorial practices in evolving alongside new technologies and contexts, the independent curator may have been instrumental in shaping contemporary curatorial discourse and challenging institutional norms. With the turn towards the meta-curator in the early 2000s (those being self-referential, and reflexive of their practice), they were never entirely free from the structures they aimed to critique, and turned the curatorial practice into a closed, self-referential loop.

Although the practice of online curating is hardly reflected in the general discourse of the curatorial, in what ways do New Scenario's antagonistic attitude to exhibition spaces move beyond the zombie curator? In most scenes in *HOPE* zombies are part of the decoration; they are not sedated but have clearly wrecked the institution. Yet in others they could be seen to reflect the institutional, as happens in the library where an army of zombies stands on top of the tables. They all look the same, reflecting the homogeneity of the military (albeit these are blue-faced females).

---

bile, flexible curatorial voices. For more information, see Paul O'Neill (2012): The Culture of Curating and the Curating of Culture(s), Cambridge, MA: MIT Press.

40 Ibid.

41 A. Dewdney: Forget Photography.

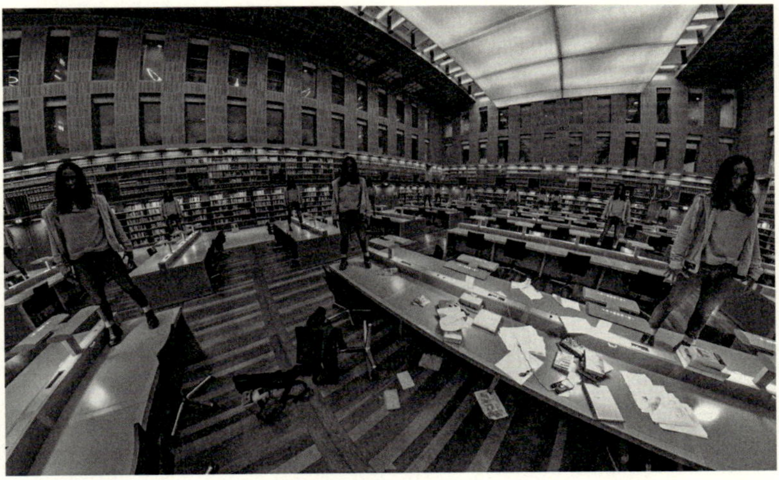

fig. 4: New Scenario: H O P E, Library, 2017

While they seem to revolt against the system, their apathetic stares and repetitive gestures are nothing like the unruly mob in the opening scene. This is an army of the same zombie, serially produced to strike one pose. Remarkably, despite a few books thrown on the floor, little is disrupted in the institutional space of knowledge. Clicking on the books opens articles by members of the commissioning organisation analysing the notion of the zombie: for instance, its relation to perpetuating colonialism, or how the zombie can be seen as an »antithesis« in which the »subject position is nullified, not reinvigorated«. According to one of the authors, Tanja Prokić, the zombie can be read

> »in terms of the pressure to innovate, the ever-shorter production cycles aimed at ever-higher profit margins that force pseudo-renewals and turn new to old at the speed of light. [...] That practically every human still alive is considered already infected and at death, however it occurs, will himself become a zombie, is the natural law of neoliberalism, by which the serial production of all products, visibly building one on the other, decide the status and relevance of their owners.«[42]

42 Tanja Prokic (2017): »Zombie Media. One Zombie Does Not An Apocalypse Make«, in: *H O P E*. Online: http://newscenario.net/hope/ (last access: 15.08.2025).

Here the political against the institutional is revealed through the notion of the zombie. In other situations this is less the case, and it could even be argued that in the presentation of the artworks – once found – world-building remains in the realm of the conventional curatorial approach and even its media usage. Beyond the disruption and despair of the zombie, essentially the spatial and even temporal condition of *H O P E* follows conventional approaches: showing pop-ups of artworks, thereby disconnecting them from their environment. The fear of becoming a zombie curator is what may have propelled New Scenario to present *H O P E*, but it is still ambiguous if the curator is freed from the institutional enclosure, or if it is only the artwork that breaks out and persists in another context. Perhaps, in this sense, and following Zizek's analysis of George Romero's film *The Night of the Living Dead*, the undead »are not portrayed as embodiments of pure evil, of a simple drive to kill or revenge, but as sufferers, pursuing their victims with an awkward persistence, colored by a kind of sadness«,[43] the curator is the one who suffers while trying to break out of the white walls, hoping to persuade the audience by constructing an extreme *mise-en-scene*.

It could be argued that, similar to the meta-curator, online curating reflects not only on *what* is shown but *how, when,* and *where* it is shown, embedding this reflection within the exhibition itself. As New Scenario demonstrates, this approach can empower radical experimentation and critical discourse; at the same time, the figure of the curator remains. Similarly, New Scenario does not practice networked co-curation,[44] nor do they use open-source methods and algorithmic systems following a post-human curatorial model.[45] Yet, they have a keen awareness of how technological environments mediate curatorial agency. Unlike conventional online exhibitions that merely replicate gallery logics in digital space, *H O P E* adopts a *site-specific* approach: produced not *in spite* of the internet, but *for* it. This mode acknowledges that exhibitions in technical, networked environments are shaped as much by infrastructures, software and platforms as they are by curatorial intention. Finally, the use

43 Slavoj Žižek (1989): »The Real and Its Vicissitudes. Return of the Living Dead«, in: Newsletter of the Freudian Field 3/1-2, pp. 80–101, here: p. 82.

44 A. Dekker/G. Tedone: Networked Co-curation, n.p.

45 M. Tyżlik-Carver: Curator | Curating | the Curatorial.

of documentation by artists has a long tradition,[46] and its significance has been further amplified in the context of preservation.[47] However, within exhibitions, documentation is often regarded as secondary, a derivative of the original artwork that may no longer be accessible.[48] Challenging this perception, New Scenario's curatorial practice offers a new perspective on the role and value of documentation. Using documentation to create a space in which artworks are presented, some of which as documentation, turns documentation into a meta-practice: it is documentation of something that happened, it is a new exhibition format which contains documentation and it is a performance that can be potentially documented. As a form of enduring documentation, something that continually folds back, it is a method to create a space where the past remains dynamic that moves beyond preserving a static memory of an event by keeping its potential alive. Finally, by presenting their exhibition in an online environment rather than displaying artworks in unconventional physical contexts, they forge additional meaning, while also showing the relevance of providing distribution and access over experiencing the original artwork. In doing so, they challenge the prevailing notion that art must be experienced in person to be fully appreciated, as well as how the copy is just as, or even more, relevant than its original.

While traditional binaries dissolve (artist vs. curator, viewer vs. artwork, object vs. process) into a dynamic, mutable system, there is also a tension between New Scenario's bold formal experimentation and a relatively underdeveloped focus on the audience's experience or the deeper dependencies of curating on web-based infrastructure. In this sense, their online exhibitions operate both within and beyond *zombie curating*: indeed, it has become a living-dead hybrid of forms and functions that challenges traditional exhibitions but also doesn't detach completely. Rather it can be seen as zigzagging between these positions: taking sharp turns that are characterised by lateral shifts across different domains rather than forward progress within a single domain. In a sense, it is similar to the notion of zombie media, which is

46 Annet Dekker (2023): »The Tension Between Static Documentation and Dynamic Digital Art«, in: Id./Gabriella Giannachi (eds.), Documentation as Art: Expanded Digital Practices, London: Routledge, Museum & Heritage Studies, pp. 16–26.

47 Annet Dekker/Gabriella Giannachi (eds.) (2023): Documentation as Art: Expanded Digital Practices, London: Routledge, Museum & Heritage Studies.

48 Barbara Clausen (2017): »Performing the Archive and Exhibiting the Ephemeral«, in: Gabriella Giannachi/Jonah Westerman (eds.), Histories of Performance Documentation, London: Routledge, pp. 93–114.

»concerned with media that is not only out of use, but resurrected to new uses, contexts and adaptations«,[49] and where minor changes can have major consequences. New Scenario also repurposes old formats for alternative uses, leading to a field of entangled positions and shifting agencies. Here, domains do not merely coexist nor do they oppose; they *shift into one another*, collapsing boundaries and inviting more layered, complex readings. *H O P E* looks like a film, a mashup of paintings by Hieronymus Bosch and Pieter Bruegel, a baroque opera or a binaural beat, functions like a game, feels like an immersive experience or a lucid dream, and reads like a visual manifesto – all at once. By resisting singular classification, it dismantles the underlying assumptions that categories in art and curating are natural, fixed, or necessary. This is not merely an aesthetic gesture, but a strategic reconfiguration of how to build, interact with and comprehend exhibitions.

The End

Through their exhibitions New Scenario rethinks how art can engage with the world. Rather than offering control, coherence, or clarity, their curated environments emphasise affective entanglement, and uncomfortable intimacy. For instance, in *Body Holes*, the artworks are placed inside mouths, ears, and other bodily orifices. It is grotesque, absurd and also relational. Artworks do not sit on pedestals, rather they inhabit, seep, and stir. The spaces are not neutral, they are fleshy, affective terrains. Nor does the viewer observe from a safe distance, they are drawn into co-presence with the world being built. *H O P E* connects to world-building in a literal sense, but the actions and methods also align closely with Haraway's insistence that worlding is always embedded in bodies, relations, and material specificity. Similar to Tsing's *matsutake* forests or Haraway's companions, New Scenario's exhibitions create spaces where meaning is emergent, dispersed, and entangled. Moreover, rather than simply building immersive environments, their practice is about imagining and inhabiting alternative ways of presenting and experiencing art, ways that are messy, unresolved, and full of potentiality. Their exhibitions also operate as narrative environments: each with its own aesthetic grammar, its own spatial and temporal rules. In an attempt to emphasise the constant negotiation between the researcher's judgement, the player's agency, the artists' visions, the curatorial

49 G. Hertz/J. Parikka: Zombie Media, p. 429.

intentions, the audience's expectations, and the dynamic nature of the environment, the three characters in this story – Observer, Player, Art/Artists-Actors-Audience – serve to encourage the inhabiting of multiple viewpoints simultaneously. Their perspectives slowly evolve and unfold, much like an allegory presents different layers that become apparent when shifting position. The Observer, restricted by their neutral stance, is limited to following the flow of the world as it has been designed. They experience a fixed, symbolic world rather than actively creating or evolving it. The Player moves in and out of different worlds, the zombie horror and reflective artworks coalesce, and are simultaneously obstructions and clues to navigate and make sense of the game. The Art is divided and connected by three protagonists Artists-Actors-Audience, who each from their own perspective, together possess a pseudo-agency; due to their constraints, they engage with the world from a distance. Both immersed and stuck in their world, they do not have the power to shape or transform it in a meaningful way. The three main characters also allude to the notion of curatorial worlding, in which the Observer provides the conventional guidance and reflection. They may be disturbed by the environment but become invigorated by the un-abstruse presentation of the art; their subjectivity emerges slowly. The Player's stimulus is to get into the gore: driven by past experiences, they understand and play with the dualities. Their incentive is to become immersed, to quickly find and connect the dots that will provide the solution. The Art/Artists-Actors-Audience consists of those who fill the space and may give it a more layered meaning in the virtual/actual. Their pseudo-agency is integral to this process, as it suggests that while they are agents within these worlds, they are still part of a larger structure, one that may invite participation but also imposes limits on what can be done, thereby creating a tension that is fundamental to worlding.

By imagining impossible geographies and sensory logics that sit outside standardised norms, New Scenario disrupts traditional curatorial methods in favour of creating spaces where the boundaries between art, artists, actors, audience, and the context itself, become porous and shift constantly. The exhibitions are not merely spaces to be visited, but fluid, unstable, and transformative environments to be inhabited. In these spaces, meaning is not static or fixed; instead, it emerges from the ongoing dialogue between different ›worlds‹. Each interaction with the exhibition, whether through the eyes of the viewer or the curatorial decisions that shape it, participates in the continuous worlding of the exhibition. This process of worlding emphasises the multiplicity of perspectives and experiences that can emerge in response to art, but also highlights that exhibitions (or artworks) are not solely defined by their visual representations or even their immediate aesthetic impact, but rather by the relational and contextual processes that unfold around or between them.

In this way, New Scenario resists the flattening of curatorial intent, and simultaneously questions authorship itself, thereby opening up the exhibition to a pluralistic, participatory process of meaning-making. Through this framework, exhibitions become more than passive displays, they become active sites of transformation where the artwork, the exhibition (including its technical, spatial and temporal qualities), and the viewer contribute to the creation of meaning in an ongoing, unpredictable interplay. Such an approach reshapes the notion of curatorial intent by recognising the dynamic, ever-transforming force of online exhibitions. Ultimately, in New Scenario's zombie landscape, nothing is dead and nothing is fully alive. These are exhibitions not to be visited, but inhabited: spaces where meaning is not delivered, but lived through, as an unstable and provocative experience. Still, the examples also emphasise a hybrid state of the curatorial. This tension is central to understanding the afterlife of curating and its zombie condition. On the one hand, it continues to reflect institutional practices through the premise of selecting and presenting art. On the other hand, it begs the question: if the curatorial role disappeared, would curating still be possible, would exhibitions be recognisable? In other words, in the end the curator is a recursively nested zombie: a self-referential condition that implies its own impossibility or contradiction – an aesthetic mode characteristic of online exhibitions.

Bibliography

Arfaoui, Meryem-Bahia (2021): »Time and the Colonial State«, in: https://thefunambulist.net (21.06.20221). Online: https://thefunambulist.net/magazine/article/time-and-the-colonial-state (last access: 15.08.2025).

Castells, Manuel (2010): The Rise of Network Society. The Information Age. Economy, Society, and Culture, Vol. 1 (second edition), Oxford: Wiley-Blackwell.

Clausen, Barbara (2017): »Performing the Archive and Exhibiting the Ephemeral«, in: Gabriella Giannachi/Jonah Westerman (eds.), Histories of Performance Documentation, London: Routledge, pp. 93–114.

Comentale, Edward P. (2017): »Zombie Race«, in: Sarah Juliet Lauro (ed.), Zombie Theory: A Reader, Minnesota: University of Minnesota Press.

Dekker, Annet (2024): »Curating Online: Art in Space and Time«, ISEA2024 Proceedings, Brisbane, online: https://isea-archives.siggraph.org/wp-content/uploads/2025/01/2024_Dekker_Curating_Online_Art_in_Space_and_Time.pdf (last access: 15.08.2025).

Dekker, Annet (2023): »The Tension Between Static Documentation and Dynamic Digital Art«, in: Annet Dekker/Gabriella Giannachi (eds.), Documentation as Art: Expanded Digital Practices, London: Routledge, Museum & Heritage Studies, pp. 16–26.

Dekker, Annet/Gabriella Giannachi (eds.) (2023): Documentation as Art: Expanded Digital Practices, London: Routledge, Museum & Heritage Studies.

Dekker, Annet (ed.) (2021): Curating Digital Art. From Presenting and Collecting Digital Art to Networked Co-Curating, Amsterdam: Valiz.

Dekker, Annet/Gaia Tedone (2019): »Networked Co-curation: An Exploration of the Sociotechnical Specificities of Online Curation«, in: Arts 8/3, n.p., online: https://www.mdpi.com/2076-0752/8/3/86 (last access: 15.08.2025).

Dewdney, Andrew (2021): Forget Photography, London: Goldsmiths Press.

Duncan, Carol (1995): Civilizing Rituals. Inside Public Art Museums, London: Routledge.

Ernst, Wolfgang (2017): »The Delayed Present: Media-Induced Tempor(e)alities & Techno-traumatic Irritations of ›the Contemporary‹«, in: Geoff Cox/Jacob Lind (eds.), The Contemporary Condition, Berlin: Sternberg Press.

Ghidini, Marialaura (2019): »Curating on the Web: the Evolution of Platforms as Spaces for Producing and Disseminating Web-based Art«, in: Arts 8/3, n.p., online: https://www.mdpi.com/2076-0752/8/3/78 (last access: 15.08.2025).

Goriunova, Olga/Matthew Fuller (2019): Bleak Joys, Minneapolis/London: The University of Minnesota Press.

Goriunova, Olga (2013): »Light Heavy Weight Curating«, in: Annet Dekker (ed.), Speculative Scenarios: Or What Will Happen to Digital Art in the (Near) Future, Eindhoven: Baltan Laboratories, pp. 25–32.

Hertz, Garnet/Jussi Parikka (2012): »Zombie Media: Circuit Bending Media Archaeology into an Art Method«, in: Leonardo 45/5, pp. 424–430.

Jenkins, Henry (2006): Convergence Culture: Where old and new media collide, New York: New York University Press.

Krämer, Sybille (2006): »The Cultural Techniques of Time Axis Manipulation: On Friedrich Kittler's Conception of Media«, in: Theory, Culture & Society 23/7-8, pp. 93–109. DOI: https://doi.org/10.1177/0263276406069885

Krysa, Joasia/Magda Tyżlik-Carver (eds.) (2025): Curating Superintelligences: Speculations on the Future of Curating, AI and Hybrid Realities, London: Open Humanities Press.

Krysa, Joasia (2014): »Can Machines Curate? Notes on History of Technology Hope and Paradoxes of Curating«, in: Y. Guasque (ed.), Digital Art: Fractures, Proliferative Preservation and Affective Dimension, 5th National Symposium of the Brazilian Association of Cyberculture Researchers, MediaLan/UFG, Funape.

Lubitz, Zoey (2020): »Art As Negotiation: Jason Hirata Interviewed. Artwork that complexifies the notion of authorship«, in: https://bombmagazine.org (18.02.2020). Online: https://bombmagazine.org/articles/2020/02/18/art-as-negotiation-jason-hirata-interviewed/ (last access: 13.10.2025).

Massumi, Brian (2002): Parables for the Virtual: Movement, Affect, Sensation, Durham: Duke University Press.

Mateos, Pierre-Alexandre/Charles Teyssou (2017): »Zombie Subjectivity: New Scenario«, in: www.moussemagazine.it/ (18.09.2017). Online: http://moussemagazine.it/zombie-subjectivity-new-scenario-2017/ (last access: 15.08.2025).

Mumford, Lewis (1934): Technics and Civilization, London: Routledge.

New Scenario (2021): »New Scenario (Paul Barsch & Tilman Hornig), 1 May 2019«, in: Annet Dekker (ed.), Curating Digital Art. From Presenting and Collecting Digital Art to Networked Co-curation, Amsterdam: Valiz, pp. 202–211.

New Scenario: »Info«, in: www.newscenario.net. Online: http://www.newscenario.net/info.html (last access: 15.08.2025).

New Scenario: »Press«, in: www.newscenario.net. Online: http://www.newscenario.net/info.html selection press (last access: 15.08.2025).

O'Doherty, Brian (1976): Inside the White Cube. The Ideology of the Gallery Space, Santa Monica, CA: The Lapis Press.

O'Neill, Paul (2012): The Culture of Curating and the Curating of Culture(s), Cambridge, MA: MIT Press.

Prokic, Tanja (2017): »Zombie Media. One Zombie Does Not An Apocalypse Make«, in: *H O P E*. Online: http://newscenario.net/hope/ (last access: 15.08.2025).

Pulliam, June Michelle/Anthony J. Fonseca (eds.) (2014): Encyclopedia of the Zombie: The Walking Dead in Popular Culture and Myth, Greenwood: Bloomsbury Publishing.

Public Affairs (2021): »Virtual reality warps your sense of time. Psychology research demonstrates the unique ›time compression‹ effect of virtual reality«, in: https://news.ucsc.edu (12.05.2021). Online: https://news.ucsc.edu/2021/05/virtual-reality-time-compression/ (last access: 15.08.2025).

Scott, Tom: »The Problem with Time & Timezones«, in: www.youtube.com (30.12.2013). Online: https://www.youtube.com/watch?v=-5wpm-gesOY (last access: 15.08.2025).

Tedone, Gaia (2017): »Co-curating with Cassini: from the Abyss of Commodification to the Exploration of Space Curation«, in: Marialaura Ghidini/Rebekah Modrak (eds.), #exstrange: a Curatorial Intervention on eBay, Ann Arbour: Maize Books, pp. 180–185.

Tyżlik-Carver, Magda (2017): »| Curator | Curating | the Curatorial | Not-Just-Art Curating: A Genealogy of Posthuman Curating«, in: Springerin, The Post-Curatorial Turn 1, online: https://www.springerin.at/en/2017/1/kuratorin-kuratieren-das-kuratorische-nicht-nur-kunst-kuratieren/ (last access: 15.08.2025).

Welch, Matt (2017): »Crisis and Infection: an Interview with New Scenario on Chaos + Positivity Ahead of Their Upcoming Project *H O P E*«, in: www.aqnb.com (01.05.2017). Online: http://www.aqnb.com/2017/05/01/crisis-and-infection-new-scenario-talk-chaos-positivity-ahead-of-theirupcoming-project-h-o-p-e/ (last access: 15.08.2025).

Wolf, Mark J.P. (2012): Building Imaginary Worlds: The Theory and History of Subcreation, London/New York: Routledge.

Žižek, Slavoj (1989): »The Real and Its Vicissitudes. Return of the Living Dead«, in: Newsletter of the Freudian Field 3/1-2, pp. 80–101.

Image Credits

fig. 1–4: © New Scenario

Rethinking Presence in the Age of Mixed Reality

Chris Salter

Introduction

Let me describe an artistic situation which evokes the uncanny. I enter into a gallery and stand before a flat panel screen positioned in portrait format and with a small camera attached on the side. The image on the screen is that of a body standing in A-Pose, a term originating from computer animation that describes a natural stance with arms angled downward at about 45 degrees, forming an A shape with the body. As I position myself into this pose, which is reflected in the whited-out figure standing before me on the screen, a sudden flash takes place and the figure vanishes. I am then greeted with the phrase ›the system is processing you. Please take a headset‹. The attendant to this strange ritual provides me with a bulky and almost wrap around headset – a pair of virtual reality glasses that are then strapped uncomfortably to my face. I am greeted by what sounds like elevator music: annoying Muzak that is occasionally interrupted with a disembodied voice stating ›please hold the line‹. I am told that the experience will begin. There is a sharp blast of white noise and suddenly I am confronted with my self-image standing before a mirror in the real space. Of course, I recognize that this apparition, a kind of double or *doppelgänger* which seems to have been generated from the capture of my body with the camera, is *fake*. It is derived from a contemporary technology – the video passthrough mode of the headset which allows a real time video feed of the physical place I am standing in to be augmented with this computationally produced double. The image both appears to be a realistic rendering of me but, at the very same moment, is riddled with errors and mistakes. There is a strange hole in the side of my arm. The grinning face which stares into the mirror, is frozen, as if in suspended animation. The feet seem to be a reflection of my feet but are distorted and twisted. Suddenly, the mirror begins to crack, accompanied by breaking noises. The white noise blast comes again, and the screen goes black.

It is admittedly a strange experience to be confronted with oneself, not only in the mirror as we would expect but also by way of a worn technological device. In Austrian psychologist Otto Rank's 1925 work on the double, the mirror itself served as the primary site for the projection of self-preservation and immortality.[1] Later, Freud would revise Rank's notion to describe what he famously termed »the uncanny« – when that which should be familiar becomes strange, unsettling and threatening. The double of oneself is not only unsettling because it is a copy. It also disturbs the psyche because it causes one to doubt the solidity of one's self.

The project I have described here, coincidentally named *doppelgänger:apparatus*, which was developed at the Immersive Arts Space, a research-creation lab based at the Zurich University of the Arts (ZHdK) which I direct, is not meant to discuss psychoanalytic understandings of the self. Instead, the description aims to provoke a different question: namely, what would *presence* signify in this context? That is, the context of wearing a virtual reality headset (in this case, Meta's Quest 3.0 VR headset) that attempts to create not a virtual experience but rather a *combination* of real and constructed. Such a *mixed reality* (MR), a term coined by computer scientists Paul Milgram and Fumio Kishino in 1994, describes ›having both‹ virtual space, on the one hand and ›reality‹ on the other, available within the same visual display environment.[2]

But the word *presence* itself is also a confusing concept with multiple meanings and contexts. Long defined as a cornerstone of *virtual reality* (VR), the technological paradigm of surrounding the user's eyes and ears with a completely computationally generated space, the concept of presence carries a specific meaning: that is, the sense of action at a distance or as Marvin Minsky termed it in the mid 1980s, ›telepresence‹.[3] In other words, as VR researcher Mel Slater has argued in his concept of the *place illusion* (informed by Minsky) one is ›present‹ in a virtual space which is not really there while ignoring the *co-extensive* real space.

Yet, since Milgram and Kishino proposed the concept of MR together with the technological development of *head mounted displays* (HMDs) through Mi-

1 Otto Rank (2012): The Double: A Psychoanalytic Study, Chapel Hill: UNC Press Books.

2 Paul Milgram/Fumio Kishino (1994): »A taxonomy of mixed reality visual displays«, in: IEICE TRANSACTIONS on Information and Systems 77/12, pp. 1321–1329.

3 Marvin Minsky (1980): »Telepresence«, in: Omni June 1980.

crosoft's Hololens in the late 1990s and then the Magic Leap in the late 2000s, such a concept of presence has been challenged by way of the emergence of a new technological paradigm: that of so-called *video passthrough*. Passthrough is a process which involves capturing video from external cameras (usually mounted on/in a device) and displaying this video image directly on a screen or the HMD in *real time*. In other words, the idea is to ›pass through‹ the outside world into the display, allowing the user to see both their physical environment as well as a computationally generated one that appears to be embedded in the real environment while wearing the device.

The argument that will unfold in this article is that such passthrough technology could suggest an expanded understanding of presence – one that departs from the idea that one is present in a virtual space where one *believes* one is located. Instead, presence and its associated term – *immersion* – now shifts from the notion of *suspension of disbelief* to that of *artifice*; that is, believing that what one experiences from the perspective of the HMD mixing up the real and the computational is not real at all but a *construction*. Because of the potential of the camera as a mediating device and the ability to embed the camera's feed with computationally generated objects through passthrough technology, there is thus a break from the idea that the mediating technology of these devices disappears and becomes transparent to the wearer. In other words, *what would it signify to think of presence and immersion within such a technological instrumentarium as the passthrough-based HMD not as engulfing the wearer's attention in a constructed world which pretends to be believable* but instead as what Walter Benjamin termed ›shock‹ – the manner in which modern technologies serve to rupture, dislocate, defamiliarize and alienate our received perception of the world?[4]

To argue this point, this article is divided into four sections and a conclusion. Section I briefly describes the context of where this argument emerges, namely, a research project appropriately called »Probing XR's Futures: Design Fiction, Bodily Experience and Critical Inquiry« which is taking place at the Immersive Arts Space, a research-creation lab based at the Zurich University of the Arts (ZHdK). This project explores the larger question of what imaginaries Extended Realities (XR) has of diverse human bodies and vice versa. Section II focuses on the core mediating material-technical *instrumentarium* of XR – the

4 Walter Benjamin (2003): »On Some Motifs in Baudelaire«, in: Howard Eiland/Michael W. Jennings (eds.), Selected Writings, 4: 1938–1940, Cambridge, MA: Belknap Press of Harvard University Press, pp. 313–355.

HMD – a device which is continually reinvented but that somehow never can escape its mediating form, despite the fact that one of the core definitions is that this device disappears; what is referred to as the illusion of non-mediation.

In Section III, I discuss the question of presence, still the holy grail of virtual reality. I explore what the basic definitions are according to the key literature and then contest them based on the emergence of so-called worn mixed reality HMDs. What comes to pass in Section IV is thus a proposition: that is, a revision of thinking around the concept of presence that has endured in VR research over the last decades. Instead, I suggest a turn to understanding the conundrum of presence in XR from another disciplinary perspective: that of theater studies and, in particular, how theater theorists have challenged the concept of presence as something akin to »being there«. The article concludes briefly then with a return to the question that was first brought up in the *doppelgänger:apparatus* project: what role can strategies play that are not based on constructing a verisimilitude of reality but instead, an experience of shock and estrangement? In other words, what kinds of aesthetic and perceptual possibilities are offered when our belief in the virtual and simulated is shattered by the effect of being caught in two spaces at once?

1. Probing XR's Futures

The aforementioned questions around presence and immersion stem from ongoing research in »Probing XR's Futures« (PXRF) a four-year research program funded by the Swiss National Science Foundation and taking place at the Immersive Arts Space at the Zurich University of the Arts (ZHdK). Within its inquiry to grasp the manner in which contemporary bodily experience transforms and is transformed through XR technologies, the project asks three core research questions: (1) How is everyday interaction in XR routinely achieved? (2) How exactly will XR change interaction and what social reciprocity and mutual access will be enabled? (3) What concrete effects and forms of discipline will be enacted on disabled bodies interacting in XR?

In addressing these questions, PXRF has three specific trajectories that form part of an experimental program of inquiry. The first, which we call the *Metabox* (in reference to behavioral psychologist B.F. Skinner's concept of the Skinner Box) involves exploring XR and its coincident space – the so-called *metaverse* – in order to ask what kind of social order is imagined by way of XR technologies and how is such a social order worked out in experimental procedures?

The second phase of the project is an artistic one, asking what kinds of future scenarios of interaction might take place in the metaverse as a social milieu? Here, we examine work in the sociology of science which is focused on ›technological promises‹ and ›social-technical imaginaries‹ where we will develop scenarios using the approach of *design fiction*. Design fiction is a critical design-oriented practice which involves creating evocative prototypes or artifacts from possible near futures to help discover new kinds of futures.[5] The kind of artifact we focus on is that of the HMD itself and what potential future scenarios can emerge through this device. For example, what happens when the HMD, which receives from its manufacturers like Meta, HTC and Samsung a yearly resolution update, doesn't actually deliver a high-resolution image of the real world but instead what Hito Steyerl has called the ›poor image‹?[6]

The third phase focuses on larger questions about the role publics play in interacting with and shaping XR technologies as well as the role of publics in transforming these technologies' already pre-packaged expectations. More specifically in the project, we plan to create different kinds of public demos: interactions with scenarios and discussions around these technologies which are normally closed off from public view within their academic research-industrial development stages. What kind of social milieu, for example, emerges when individuals start wearing these devices within public spaces: walking on the street, in the park, sitting in restaurants and bars. Indeed, as Silicon Valley imaginaries, embodied by the like of Apple's Vision Pro HMD, describe future scenarios of 24/7 embedding within these devices, the question of what kind of experiences such technologies actually propose needs to come to the fore of attention.

One of the central issues that arises within such a program of exploration around the human body as a tool for critical inquiry is what kind of bodies are imagined by the HMD. In other words, an underlying question for the research is interrogating the manner in which XR technologies like the HMD severely institute certain body norms, aiming for what Campbell has called a *corporeal*

---

5 Julian Bleeker (2009): »Design fiction. A short essay on design, science, fact and fiction«, in: Near Future laboratory, March online: https://systemsorienteddesign.net/wp-content/uploads/2011/01/DesignFiction_WebEdition.pdf (last access: 15.11.2025).

6 Hito Steyerl (2009): »In Defense of the Poor Image«, in: *e-flux journal* 10/11, pp. 1–9.

standard, an »ideal (i.e., nondisabled) body«.[7] The minor adjustments to the technologies incorporated for inclusion of people with disabilities therefore reify what disability scholar Ellcessor calls the »common sense idea of accessibility« – the perception of disability as a problem that is easily solvable by bare minimum engineering functionalities.[8] In collaboration with German disabled researcher/technologist Christian Bayerlein and PhD researcher Puneet Jain we are utilizing the mouth as an interface as well as an access point to navigation in XR. Thereby we explore how different bodily imaginaries could shape XR technology in a profoundly different way than the predominant ableist imaginary these technologies now harbour.

2. XR Instrumentarium: The HMD

The research program that has been described above would not be possible without XR's essential technological development: the head-mounted display or HMD. Despite historical arguments about the HMD either being inspired by science fiction stories or emerging from 1990s attempts to bring VR technologies to video gaming, like many Cold War technologies, the core origin stories trace back to US military contexts. Inspired by applied military research taking place at the Texas-based Bell Helicopter corporation during the 1960s, engineer and one of the founders of the field of computer graphics Ivan Sutherland developed one of the first working HMDs in 1965. At the time at Bell, the defense contractor utilized an experimental prototype display that was strapped around pilots' heads and equipped with a head tracking sensor and small CRT displays in front of the eyes and connected to remote control infrared cameras.[9]

This curious device was rumored to have been developed in order to facilitate the landing of helicopters in darkness. As the pilot turned their head, the camera turned with it. In this sense, XR, like most modern electronic and computing technology, was rooted in the midst of war, in this case, the Vietnam

7 Fiona A. K. Campbell (2001): »Inciting legal fictions. Disability's date with ontology and the ableist body of the law«, in: Griffith Law Review 10/1, pp. 42–62.

8 Elizabeth Ellcessor (2016): Restricted Access. Media, Disability, and the Politics of Participation. New York: NYU Press.

9 SIGGRAPH Conferences (2017): »VR @ 50: Ivan Sutherland's 1968 Head-Mounted 3D Display System«, in: suggraoh.org (14.08.2018). Online: https://blog.siggraph.org/2018/08/vr-at-50-celebrating-ivan-sutherland.html/ (last access: 31.08.2025).

conflict where landing helicopters in dense jungles would have clearly benefitted from the technological advance of seeing at a distance. Inspired by Bell's research, Sutherland himself set out to improve on their device, envisaging a ›kinesthetic display‹ that could »serve as many senses as possible« by inserting a computer in place of the camera to »enable the engineer to view a mathematical wonderland of their own choosing«.[10]

In its context as a kinesthetic interface to an abstract computational world, it is important to position the HMD as a *worn* form of augmentation; one »which involves the coupling of humans with computational devices, taken to extend the body's native capacities through information processing«.[11] In this sense, the HMD is not simply an extension but a creator of a »symbiotic relationship between person and device«.[12] This emphasis on the primacy of what Mark Hansen has called the *sensorimotor body* for embodied action, in which »material structures are interpenetrated with informational patterns and lay bare the enabling constraints of the body«[13] characterizes XR as a ›perceptual technology‹.[14] Such a technology serves to model human perceptual mechanisms and traits.

As such a perceptual technology, the HMD would thus evolve, oscillating between the military, the research sector and the gaming industries and finding its way into the arts by the late 1980s. For example, before the general public had ever heard of the HMD, the French visual artist Nicole Stenger, ensconced as an artist in residence at MIT's Center for Advanced Visual Studies (CAVS), developed one of the first VR-based artistic experiences. In Stenger's »virtual reality film« *Angels*, wearers of Jaron Lanier's ultra expensive VPL Eye Phone HMD and accompanying data glove interface would navigate a ›travelling in paradise‹ experience. By moving through an emotional world of sight, sound and (ultimately, non-interactive) touch, the users were to experience

---

10 Ibid.

11 Ana Viseu/Lucy Suchman (2010): »Wearable Augmentations«, in: J. Edwards/P. Harvey/P. Wade (eds.), Technologized Images, Technologized Bodies: anthropological approaches to a new politics of vision, Oxford/New York: Berghahn Books.

12 J. C. R. Licklider (1960): »Man-Computer Symbiosis«, in: IRE Transactions on Human Factors in Electronics HFE-1/1 (March), pp. 4–11.

13 Mark B. Hansen (2006): Bodies in Code: Interfaces with Digital Media, New York: Routledge, p. 22.

14 Alexander Waibel et al. (2009): »Computers in the Human Interaction Loop«, in: Id./Rainer Stiefelhagen (eds.), Computers in the Human Interaction Loop. Human–Computer Interaction Series, London: Springer, pp. 3–6.

what Stenger in her essay »Mind as a Leaking Rainbow« later described as a »new nature of reality in our everyday life [...] opening up an infinity of space in an eternity of light«.[15]

Not ironically, Stenger's multi-modal methodology of creating an experience combining sound, image and touch was also inspired by a military project: an HMD that was part of an US Air Force research program called *Super Cockpit* led by the engineer and HMD pioneer Tom Furness. For Stenger, Super Cockpit's bulky and oversized HMD represented nothing other than the dissolution of the self's solidity in the framework of a cybernetic device enrolling pilot and machine in an embodied and multi-sensory feedback loop. Such a device would »redistribute information according to the parallel tracks of vision, hearing, and touch, when the pilot faces a paralysis of decision in front of the maze of graphics and commands. At the speed of war, Descartes' world becomes brittle. Cyberspace will shatter it like a mirror«.[16]

Today, more than half a decade later, a similar mix of visionary promises, and technical specifications animates the XR industry. Augmented reality-(AR) or VR-labeled technology has been brought to broader public attention through recent IT industry innovations, rhetorical and infrastructural. These advances not only include Meta's (ex-Facebook's) now forgotten media ›Metaverse‹ and its associated technologies (e.g., Oculus Quest headsets), as well as competitor products (e.g., Apple Vision Pro). For Meta, the dream of accessing user's bodies and dreams continues the long developing cybernetic project of the HMD. For example, as of 2023, Meta had spent some 50 billion USD on research towards these displays – as the company staked their future on the Metaverse and the technologies that would enable access to it.

In following the tradition of the HMD not only as a tool for manufacturing Stenger's ›paradisical cyberspace‹ but also for pursuing the fusion of human and machine necessary for future war, Meta signed a contract in 2025 with defense contractor Anduril to develop MR headsets stocked up with AI for the US Military in order to provide soldiers with ›enhanced perception‹ and to transform them into ›technomancers.‹[17] Anduril is special in Meta's history.

15 Nicole Stenger (1991): »Mind is a Leaking Rainbow«, in: Miachel Benedik (ed.), Cyberspace: First Steps, Cambridge, MA: MIT Press, p. 51.

16 Ibid.

17 Steffan Katte (2023): »Meta to make AI-powered mixed-reality headsets for US military«, in: https://cointelegraph.com (02.06.2025). Online: https://cointelegraph.com/news/meta-ai-powered-headsets-us-military (last access: 31.08.2025).

The company was launched by Palmer Lucky, the right-wing founder of Oculus and inventor of the Oculus Rift HMD which Meta acquired for some 2 billion USD.

As HMDs have advanced from bulky, expensive and low-resolution headsets to inspirations for new forms of military machine awareness, one of the more interesting technologies that has emerged, particularly in respect to the Meta Quest is that of video passthrough. According to a recent article by a number of well-respected XR experts, including Stanford VR researcher Jeremy Balienson, while using passthrough, which is based on the cameras in the HMD, a person »does not see light from the real world but instead relies on stereoscopic, color, high resolution, low latency, real-time video of the world, which is displayed on small screens inside a headset«.[18] This observation is in contrast to light-based arrays, the main technology behind AR glasses like the *Hololens* or *Magic Leap*. Originally, developed as a safety feature to prevent users from going beyond the ›guardian‹, the sensed boundary inside an HMD display which is enclosed by a virtual fence, passthrough in such HMDs as the Meta Quest 3 or the Apple Vision Pro, has slowly moved from a safety feature to a new kind of perceptual system: a core portal into the world of MR.

As the authors of this aforementioned survey on passthrough systems make clear, there is a dearth of literature and studies around the technology. Indeed, even if Milgram and Kishino's 1994 article introducing mixed reality claimed that the paradigm did not describe a specific kind of technology but rather a taxonomic paradigm depicting a continuum between the purely virtual (i.e., computationally simulated) and the real, physical world, at the time, there were few material examples of such technologies. As Viseu and Suchman make clear, the HMD as a portal into a mixed space constructed through a computational, physical, and perceptual imaginary, is nothing less than a new form of bodily extension.[19]

18 Jeremy N. Bailenson/Brian Beams/James Brown/Cyan DeVeaux/Eugy Han/Anna CM Queiroz/Rabindra Ratan et al. (2024): »Seeing the world through digital prisms. Psychological implications of passthrough video usage in mixed reality«, in: Technology, Mind and Behavior 5/2, pp. 1–16, https://doi.org/10.1037/tmb0000129.

19 A. Viseu/L. Suchman: Wearable Augmentations, p. 62.

3. Co-Extensive Space and its Presence Effects

Passthrough-based technology might not only enable a different kind of technical imaginary. It might also shift another long-standing concept that has dominated XR-based research for decades: that is, the notion of presence. To understand what is at stake here, let's return to the artistic project that led off this article: *doppelgänger:apparatus*. According to the co-creator, Immersive Arts Space research associate Chris Elvis Leisi, the participative installation focuses on the appearance of virtual doppelgängers in MR, which act autonomously beyond the users' control. As the installation is described as one in which the virtual doppelgängers influence participants' experience of presence, social engagement and embodiment inside MR, the interesting questions that the project catalyzes is what constitutes physically embodied space that is computationally entangled and, importantly, what or who inhabits such a space. As Saker and Frith articulate, in the predominant tradition of VR, which is principally based on the HMD technology discussed in Section II, there has traditionally been little interest in what they call ›co-extensive‹ space – the developing »symbiotic relationship between physical and digital that is increasingly proximate, extensive and transformative«.[20] Yet, with the introduction of passthrough technology, »users can seamlessly shift their display between the digitality of VR and the concrete reality of their surroundings«.

Saker and Thrift argue that the vast majority of VR literature is not focused on participants' embedding in concrete space. Thus, passthrough technology offers a new possibility to go beyond VR systems ›simulation‹ of physicality that »feigns the materiality of actual space, even though the spaces mediated through VR are not concrete per se«.[21] What VR systems that cut off the real physical world from the participant do is to create »a form of presence that is outwardly dislocated from its physical setting.« In other words, what *presence* here indicates is that the user feels they are physically there – that is, present – even though their bodies are usually immobilized and effectively »transported to a dislocated space that is visually and audibly distinct from the space outside of their headset«.[22]

20 Michael Saker/Jordan Frith (2020): »Coextensive space. virtual reality and the developing relationship between the body, the digital and physical space«, in: Media, Culture & Society 42/7-8, p. 1429.

21 Ibid.

22 Ibid., p. 1430.

The mediating ability and power of passthrough technology on the HMD strongly suggests that the sense of *presence* of oneself and others (human or more than human) may actually be reconfigured through technologies of simulation that engage with the placeness of the real world. This transformation of presence thus could shift from the classic understanding of presence as ›being there‹ – what Mel Slater famously termed ›the place illusion‹. The place illusion is exactly that; a paradox that constitutes a kind of double bind which is the experience of being in a place (even though the real place one is in is usually irrelevant to the virtual experience) »in spite of the sure knowledge that you are not there«.[23] What passthrough technology adds to this picture is to complexify presence even further: that is, to see oneself and one's own body along with others in a real physical surrounding through the live camera feed of passthrough. In this sense, one is indeed in a real place, even though the perception of that place is brought to the perceiver through a mediating technology.

It seems, however, that one cannot escape the term *presence* and that its definitions, particularly in the context of psychology which has held a strong grip on VR research for some 40 years, still holds. For example, in work from Murphy and Skarbez from 2023 exploring the continued debate around presence, the researchers summarized some 30 years of VR research and deduced three core characteristics.[24] Presence is defined first and foremost by: (1) the sense of being there in the non-co-extensive space that constitutes VR-based environments; (2) the illusion of non-mediation, which suggests that users forget about the mediating power of the HMD technology; and (3), the willing suspension of disbelief – the »commitment of the person's entire neurology to the ›suspension of disbelief‹ that they are somewhere else [,] rather than where their [. . .] body really is«.[25]

Yet, these concepts of presence, particular the place illusion, which still holds as the gold standard in VR research, are based on a *historically contingent concept* – that space cannot be co-extensive because one in VR always in-

23 Mel Slater (2009): »Place illusion and plausibility can lead to realistic behaviour in immersive virtual environments«, in: Philosophical Transactions of the Royal Society B: Biological Sciences 364/1535, pp. 3549–3557.

24 Dooley Murphy/Richard Skarbez (2020): »What do we mean when we say ›presence‹?«, in: PRESENCE: Virtual and Augmented Reality 29, pp. 171–190.

25 Mel Slater/Martin Usoh (1993): »Representations systems, perceptual position, and presence in immersive virtual environments«, in: Presence: Teleoperators & Virtual Environments 2/3, pp. 221–233.

habits a computational (in the HMD) and a real space (outside of it). This concept of presence harkens back to American AI pioneer Marvin Minsky's notion of *telepresence* – which Minsky in the original context specifically describes as action at a distance and more importantly, the control of a robot in one location from another location (this was part of Minsky's vision for a telepresence-based economy). Minsky's description emerges from a specific historical period in the United States in the mid 1980s around the Three Mile Island nuclear disaster and so the idea that someone could control a machine to inspect damage and repair a reactor was appealing. Minsky states that »the biggest challenge to developing telepresence is achieving that sense of ›being there‹. Can telepresence be a true substitute for the real thing? Will we be able to couple our artificial devices naturally and comfortably to work together with the sensory mechanisms of human organisms?«[26] Thus, a historically contingent idea – the concept of ›being there‹ – in the case of VR situated in a computationally generated space, was transformed into scientific criteria and the basis of a definition (that of presence) that seems to hold, regardless of the technological development of the HMD.

We can explore how this concept of presence as ›being there‹ starts to be challenged by passthrough by briefly examining another artistic project, one that the author of this chapter developed between 2021–2023. *Animate* is a theater work that utilizes the techniques of XR to tell the story of a doomed relationship in a near future Canada radically transformed by climate change. Premiered in August 2022 at the Kunstfest Weimar performing arts festival in Germany and divided into three acts, *Animate* resonates specifically with researcher Ronald Azuma's argument of focusing on »virtual content that is integrated with the surrounding real world, while users remain engaged with and aware of that ›real world‹«.[27] In the three act story, the audience moves between three genres: (1) live theater that is unmediated by digital technologies; (2) VR, in which the audience wears VR headsets and is collectively led by the actors through a physical space; and (3) an MR finale in which the audience encounters the earth coming alive in an apocalyptic scene of animated 3-D flying rocks that are mapped into the real space and which visually and acoustically attack both characters and audience.

26 M. Minsky: Telepresence.

27 Ronald T. Azuma (2006): »The most important challenge facing augmented reality«, in: Presence 25/3, pp. 234–238.

What is salient about the MR setting in *Animate*'s third act is the production of a *believable* experience – animated rocks of varying shapes and sizes and with specific physics that explode over, chase and in general, produce the feeling of a climate change-oriented apocalyptic end game in the audience. Interestingly enough, even though the audience wearing the HMDs knows that there are no real (i.e., physically present) flying rocks in the physical space, they act as if the rocks were actually there. This believability aligns with a renewed notion of presence proposed by Skarbez in which presence is defined as »the perceived realness of a mediated or virtual experience«.[28] According to the authors, realness is perceived in contrast to actual realness as a »function of a system's ability to provide stimuli that match reality«.[29]

But this so-called Skarbez model also incorporates another key trope which is the complement to presence in VR research: the so-called *plausibility illusion*. According to Slater, the plausibility illusion (or Psi) »is the illusion that what is apparently happening is really happening (even though you know for sure that it is not)«.[30] Psi involves what Slater defines as »correlations between external events not directly caused by the participant and his/her own sensations (both exteroceptive and interoceptive)«.[31] Psi is based on the virtual environment reacting to the actions of the user, events that coherently incorporate the user into the action and finally, a kind of verisimilitude that the simulation could happen in reality.

More recently and in light of increased research in MR, researchers are attempting to revisit these conceptions of presence. For example, Skarbez introduces an additional argument, namely, that plausibility should be augmented by the notion of *coherence*: »the extent to which a virtual environment behaves in a reasonable or predictable way«.[32] In another article, a group of researchers led by Slater argue that while the place illusion is ›inverted‹ in AR (incorporating virtual objects into the real world) the plausibility illusion stays the same:

28 Richard Skarbez/Frederick P. Brooks Jr./Mary C. Whitton (2017): »A survey of presence and related concepts«, in: ACM computing surveys (CSUR) 50/6, pp.1–39.

29 Ibid.

30 M. Slater: Place illusion, p. 3550.

31 Ibid., p. 3551.

32 Richard Skarbez/Frederick P. Brooks/Mary C. Whitton (2020): »Immersion and coherence: Research agenda and early results«, in: IEEE transactions on visualization and computer graphics 27/10, p. 3840.

the virtual objects that are embedded into this real world must be seen as real.[33] The researchers point to how this might be done, for example, through the casting of virtual shadows on real objects or the occlusion of real objects with virtual ones.

At first, one could attribute *Animate's* audience running from virtual rocks as exemplary of Psi. But the aesthetic experience of rocks exploding in one's midst and then forming a chain that rushes through a space certainly is not a simulation of what happens in lived reality. The fact that *Animate* utilized an early version of Meta's passthrough control (not an API) meant that the system would occasionally result in glitches, for example, where the virtual rocks would not occlude real objects in the space or when suddenly, the same virtual rocks would ›bleed‹ through the real visitors in the video feed. Indeed, one would think from the scientific definitions of Psi that belief in the virtual/real world overlay would shatter through such glitches and that the audience would stand in place while 3-D photorealistic rocks would move towards them at accelerated speeds. But this was not the case. So where does this leave Psi and place illusion in trying to understand the role that immersion plays in MR experiences?

4. Revising Belief

So far, we have explored how the emergence of a new technological instrumentarium, the video passthrough augmented HMD, begins to question long established definitions of immersion and presence that describe HMD based AR. The question that remains, however, is not whether the sense of presence is better or worse, or higher over lower fidelity in MR or VR. Indeed, in our *doppelgänger:apparatus* example, the captured and reconstructed image of the visitor, is certainly not a perfect replica or copy. It is marked with digitally generated glitches due to technical issues around the quality and control of ambient light, the background setting in which the visitor stands and even the color of their clothing. Rather, the question is how presence might be reformulated not only due to a new kind of mediating technology (passthrough based HMD) but also a new phenomenology of this technology *in use*. This notion thus takes into

33 Mel Slater/Donma Banakou/Alejandro Beacco/Jaime Gallego/Francisco Macia-Varela/Ramon Oliva (2022): »A separate reality. An update on place illusion and plausibility in virtual reality«, in: Frontiers in virtual reality 3/914392.

account the tension between the perceptual experience of being in one's body inside the physical co-extensive space while having that space mediated by a visual sensing technology – that is, the camera.

Of course, even though there is a form of technical mediation taking place, one would find it challenging to argue that the space one is seeing with the camera – even though its live feed might be subjected to visual distortions, frame drop outs, lower pixel resolution, and the like – is *not the same* co-extensive space that one moves within. This is because of a host of *sensorimotor contingencies*, the idea that the different senses give us multiple channels of feedback about where our bodies are in space and in time.[34] In other words, what I have argued is that what passthrough might do is create a tension between the co-extensive and non co-extensive space – as potentially seen in the *doppelgänger:apparatus* and *Animate* examples. But how then does the concept of presence get rethought through this tension?

One of the key ways that immersion and presence have been defined in the VR literature derives from the idea of the suspension of disbelief (sod). Although the notion originally can be traced to the poet Samuel Taylor Coleridge, sod is also used frequently in another disciplinary context: that of theatre and performance theory. Here, sod describes where we »will ourselves to ›see as,‹ without the necessity of belief«.[35] In other words, we suspend the belief that what we are watching in front of us (usually, in the theater's case, on the stage) is *not* real and take on the stance that we will *believe* that the artificial is real. Clearly, presence theories in VR are built on this cornerstone: that the interactor wearing the HMD believes that their body spatially inhabits a ›real‹ space (it is self-located there because the wearer of the HMD always has the 1st Person Point of View) even though they know that what they are seeing (and sometimes hearing) is artificially constructed.

But isn't this the same as with a theatrical performance? Isn't it also true that the actors (in the context of dramatic theatre) are playing characters that are also not present and that the audience's bodies are simultaneously in both a fictional space and also a real one? If one doubts this, just imagine a typical theatre experience when the real intrudes on the fictional as your limbs fall asleep during a poor performance. In this sense, you inhabit the place illusion,

---

34 J. Kevin O'Regan/Alva Noë (2001): »A sensorimotor account of vision and visual consciousness«, in: Behavioral and brain sciences 24/5, pp. 939–973.

35 Marvin Carlson (1989): »The iconic stage«, in: Journal of Dramatic Theory and Criticism 3/2, pp. 3–18, here: p. 4.

oscillating between the real space and, at the same time, a virtual one – that of the stage event. In other words, as the theatre studies scholar Cormac Power states, theatre is an overlapping of illusion and reality. Similarly, applied to the passthrough experience, the participant overlaps both spaces. In this way, we can say that the passthrough based MR experience »can be seen not so much as ›having‹ or containing (or being defined as) presence but rather *functioning to play with its possibilities*«. In other words, the passthrough technology creates a strange perceptual disjunction between our awareness of artifice (we know the digitally produced objects or, in the case of the *doppelgänger:apparatus* project, the participant's virtual double) and the felt and seen reality of our very present bodies in the actual space, by way of the camera feed.[36]

This process of suspension of disbelief, of believing you are in one space while your body is in another, is not unique to VR. It is also highly emblematic of dramatic theatre where characters are represented on the stage but at the same time are absent. In drawing on the vast literature from Derrida to the theatre phenomenologist Bert O' States that problematizes theatrical theories of presence, such a theory assumes that the theatre is somehow from the start an *experience that one believes is real and that presence is based on such a belief*.

Yet, we need to remind ourselves that the theatre, like other art forms is, in essence, an artifice and construction. This is where we thus come full circle to our question of how the HMD as a technological device mediates these conceptions of presence and immersion. Indeed, if the suspension of belief model partially connects theatre to older ideas of being there in a virtual world, what it leaves out is another key element of the stage: namely its *technicity* that is made possible from viewing the stage itself as a technical system. As theatre theorists argue, one key notion of presence in performance is that of the literal fact of ›being-there‹ in-the-same-room-at-the-same-time as the artist/character. This kind of being there at once implies both a rejection of a transcendent stage ›aura‹ (that the stage is somehow more present) and an eschewal of attempts to create theatrical illusion.

Thus, we could argue that the concept of immersion might not then be defined by the belief of being there but by a *break*, a rupture and shock that the world that comes to the wearer of the HMD is *anything but real*. Such a concept is certainly not new. In fact, it broadly defines the 20th century theatre avant-garde where artists from Meyerhold to Brecht attempted to tear down

36 Cormac Power (2008): Presence in play. A critique of theories of presence in the theatre, Amsterdam: Rodopi.

the conception of the theatre as a realistic experience, instead creating works that aimed to deconstruct the expectation of real life portrayed on the stage. In fact, Power argues that a key feature of theatrical experience involves the simultaneity of imaginatively ›seeing‹ a fictional world that has been conjured up, while also seeing the theatrical means/machinery of creating that fictional world. Likewise, theater's technicity »both stages and represents the technologies it dramatizes, including those apparently immediate to theater«.[37] Such a conception of performance as expressive of technicity could also be said to apply to the fundamental technical mediation of perception that takes place within the realm of XR. This conception also radically departs from the concept that VR's understanding of presence is also determined by the illusion of non-mediation (forgetting that one is wearing a technical device) functioning for users.

Conclusion

The break from immersion as suspension of disbelief also harkens back to notions emerging from Russian formalism in the early 1920s about thinking poetry (or art in general) as a device to understand the ›literariness‹ of literature or ›the stoniness of a stone‹. Power argues something similar in his position that theater's basis in the world is that which allows us to perceive its *materiality* in a different way – perhaps even a strange manner. Thus, in the context of aesthetic experience, it becomes necessary to reposition the manner in which XR is thought. That is, to move away from the idea that transparency and non-mediation is the guarantee of immersion. Instead, immersion is really about fracturing belief – shocking the perceiver through technical mediation to begin to see the world not as it is given but in its artifice as a social-technical construction. Perhaps herein thus lies the political potential of immersion as shock – to unmask the ideologies that drive technologies like mixed reality by defamiliarizing our use and experience of these systems.

Here, the technologies harnessed to construct immersion function as in what the Russian literary critic Viktor Shlovsky described as a *device*. In a now influential 1924 essay entitled »Art as Device«, Shlovsky argued that the aim of poetry is to make the reader aware of the devices that make people aware.

37 William B. Worthen (2020): Shakespeare, Technicity, Theatre, Cambridge UK: Cambridge University Press.

This conception of estrangement (*ostranie*) is traditionally not discussed in the context of immersion. Yet, estrangement in Shlovksy's and later Brecht's context of *verfremdung* (defamiliarization) suggests that performative immersion through distance and shock is actually not in opposition but goes hand in hand. In other words, immersion is not belief in a world which is ultimately not there. Rather, through the very shock of technical mediation, immersion brings us into direct confrontation with what Shlovsky said defined art: the very fact that »the process of perception is, in art, an end in itself and must be prolonged. Art is a way of experiencing the artfulness of an object.«[38]

Bibliography

Azuma, Ronald T. (2006): »The most important challenge facing augmented reality«, in: Presence 25/3, pp. 234–238.

Bailenson, Jeremy N./Beams, Brian/Brown, James/DeVeaux, Cyan/Han, Eugy/Queiroz, Anna CM/Ratan, Rabindra et al. (2024): »Seeing the world through digital prisms. Psychological implications of passthrough video usage in mixed reality«, in: Technology, Mind and Behavior 5/2, pp. 1–16. DOI: https://doi.org/10.1037/tmb0000129

Benjamin, Walter (2003): »On Some Motifs in Baudelaire«, in: Howard Eiland/Michael W. Jennings (eds.), Selected Writings, 4: 1938–1940, Cambridge, MA: Belknap Press of Harvard University Press, pp. 313–355.

Bleeker, Julian (2009): »Design fiction. A short essay on design, science, fact and fiction«, in: Near Future laboratory March, online: https://systemsorienteddesign.net/wp-content/uploads/2011/01/DesignFiction_WebEdition.pdf (last access: 15.11.2025).

Campbell, Fiona A. K. (2001): »Inciting legal fictions. Disability's date with ontology and the ableist body of the law«, in: Griffith Law Review 10/1, pp. 42–62.

Carlson, Marvin (1989): »The iconic stage«, in: Journal of Dramatic Theory and Criticism 3/2, pp. 3–18.

Ellcessor, Elizabeth (2016): Restricted Access. Media, Disability, and the Politics of Participation, New York: NYU Press.

38 Victor Shklovsky (1965): »Art as Technique«, in: L. Lemon/M. Reis (eds.), Russian Formalist Criticism, Nebraska: University of Nebraska Press, pp. 3–24, here: p. 12.

Hansen, Mark B. (2006): Bodies in Code. Interfaces with Digital Media, New York: Routledge.

Katte, Stefan (2023): »Meta to make AI-powered mixed-reality headsets for US military«, in: cointelegraph.com (02.06.2025). Online: https://cointelegraph.com/news/meta-ai-powered-headsets-us-military (last access: 31.08.2025).

Licklider, J. C. R. (1960): »Man-Computer Symbiosis«, in: IRE Transactions on Human Factors in Electronics HFE-1/1 (March), pp. 4–11.

Milgram, Paul/Fumio Kishino (1994): »A taxonomy of mixed reality visual displays«, in: IEICE TRANSACTIONS on Information and Systems 77/12, pp. 1321–1329.

Minsky, Marvin (1980): »Telepresence«, in: Omni June 1980.

Murphy, Dooley/Richard Skarbez (2020): »What do we mean when we say ›presence‹?«, in: PRESENCE: Virtual and Augmented Reality 29, pp. 171–190.

O'Regan, J. Kevin/Alva Noë (2001): »A sensorimotor account of vision and visual consciousness«, in: Behavioral and brain sciences 24/5, pp. 939–973.

Power, Cormac (2008): Presence in play. A critique of theories of presence in the theatre, Amsterdam: Rodopi.

Saker, Michael/Jordan Frith (2020): »Coextensive space. virtual reality and the developing relationship between the body, the digital and physical space«, in: Media, Culture & Society 42/7-8, pp. 1427–1442.

Shklovsky, Victor (1965): »Art as Technique«, in: L. Lemon/M. Reis (eds.), Russian Formalist Criticism, Nebraska: University of Nebraska Press, pp. 3–24.

SIGGRAPH Conferences (2017): »VR @ 50: Ivan Sutherland's 1968 Head-Mounted 3D Display System«, in: siggraph.org (14.08.2018). Online: https://blog.siggraph.org/2018/08/vr-at-50-celebrating-ivan-sutherland.html/ (last access: 31.08.2025).

Skarbez, Richard/Frederick P. Brooks/Mary C. Whitton (2017): »A survey of presence and related concepts«, in: ACM computing surveys (CSUR) 50/6, pp. 1–39.

Skarbez, Richard/Frederick P. Brooks/Mary C. Whitton (2020): »Immersion and coherence: Research agenda and early results«, in: IEEE transactions on visualization and computer graphics 27/10, pp. 3839–3850.

Slater, Mel (2009): »Place illusion and plausibility can lead to realistic behaviour in immersive virtual environments«, in: Philosophical Transactions of the Royal Society B: Biological Sciences 364/1535, pp. 3549–3557.

Slater, Mel/Usoh, Martin (1993): »Representations systems, perceptual position, and presence in immersive virtual environments«, in: Presence: Teleoperators & Virtual Environments 2/3, pp. 221–233.

Slater, Mel/Domna Banakou/Alejandro Beacco/Jaime Gallego/Francisco Macia-Varela/Ramon Oliva (2022): »A separate reality. An update on place illusion and plausibility in virtual reality«, in: Frontiers in virtual reality 3/914392.

Stenger, Nicole (1991): »Mind is a Leaking Rainbow«, in: Michael Benedik (ed.), Cyberspace: First Steps, Cambridge, MA: MIT Press, pp. 49–58.

Steyerl, Hito (2009): »In Defense of the Poor Image«, in: e-flux journal 10/11, pp. 1–9.

Sutherland, Ivan E. (1965): »The ultimate display«, in: Proceedings of the IFIP Congress 2, pp. 506–508.

Viseu, Ana/Lucy Suchman (2010): »Wearable Augmentations«, in: J. Edwards/P. Harvey/P. Wade (eds.), Technologized Images, Technologized Bodies: anthropological approaches to a new politics of vision, Oxford/New York: Berghahn Books.

Waibel, Alexander et al. (2009): »Computers in the Human Interaction Loop«, in: Alexander Waibel, Rainer Stiefelhagen (eds.), Computers in the Human Interaction Loop, Human–Computer Interaction Series, London: Springer, pp. 3–6.

Worthen, William B. (2020): Shakespeare, technicity, theatre, Cambridge UK: Cambridge University Press.

EVERYDAY LIFE/ART

Modes of Reception in VR

Simulation as Attunement

Metastability, Coherence, and Embodied Orientation in Virtual Reality

Jose Hopkins B., Ksenia Fedorova

As immersive technologies continue to change, the question of what it means to *feel real* within virtual systems becomes increasingly urgent. Users often report moments of alignment, recognition, and clarity while engaging with virtual environments that are unstable and materially disjointed. This paradox lies at the heart of our inquiry: how does embodied coherence emerge under conditions of sensory and spatial instability?

We build on the argument that VR, MR and XR[1] do not function by delivering a total illusion,[2] but by enabling the body to *attune* to shifting condi-

1 We use the terms VR, MR and XR interchangeably throughout this paper. We echo Mark B. Hansen's claim that all virtual reality is in fact »mixed reality« because it is always experienced by and through the body, which always remains partially »outside«. As we focus on embodied processes rather than discrete technological categories, we decided to continue building on the theoretical current that argues that VR experiences are always situated in »real« physical space, and thus are always mixed reality (MR). See Mark B. Hansen (2006): Bodies in Code. Interfaces with Digital Media, New York: Routledge; and Liam Jarvis (2021): Immersive Embodiment. Theatres of Mislocalized Sensation, London: Palgrave Macmillan.

2 By »total illusion«, we refer to the idea that immersive systems (such as VR) aim to completely replace reality with a seamless, perceptually convincing simulation that the user accepts as continuous and self-contained. This view aligns with early »presence« theories in VR research, where success was measured by the degree to which users »forgot« the real world (Cf. Mel Slater/Sylvia Wilbur (1997): »A Framework for Immersive Virtual Environments (FIVE): Speculations on the Role of Presence in Virtual Environments«, in: Presence: Teleoperators and Virtual Environments 6/6, pp. 603–616). Our approach challenges this notion by emphasising that XR experiences are not total illusions to be believed, but modulations to be navigated. Even in the most »immersive« environments, the body remains partially outside simulation – proprioceptively,

tions, offering an opportunity to reconfigure one's inner sense of the self and to *modulate* its *coherence* within a fluctuating perceptual field. *Simulation*, in this context, is not a seamless reproduction or a model of reality – not a stable representational layer that precedes experience. It is a dynamic, co-emergent field of interaction, meaningful only through embodied engagement. More than an illusion, simulation functions as a regulatory force: a structure of sensory and affective modulation through which the body learns to feel itself aligned, even if only momentarily. The body is never entirely »inside« the simulation; it acts from within and without, participating in what we call *metastable interfacing* – a continuous oscillation between the actual and the virtual. By *metastability*, we do not refer to a property of the digital system or environment, but to the creative and responsive field that emerges through affective and proprioceptive modulation. It is not the system that is metastable, but the relation: a constantly negotiated state of readiness, reorientation and becoming.

Our central claim is that simulation operates as *attunement*: it does not mask instability but guides the body through it. In doing so, it enables coherence not as a given state, but as a felt achievement; enacted through orientation, recalibration, and movement across the shifting seams of XR experience.

1. Metastable Interfaces: Embodiment in Tension

Across multiple disciplines, there is growing acknowledgement of the body's flexibility and its relational openness to its environment. Thinkers in phenomenology (Maurice Merleau-Ponty), process philosophy (Erin Manning, Brian Massumi, Gilles Deleuze), embodied cognition (Francisco Varela, Alva Noë, Shaun Gallagher) post-phenomenology (Don Ihde, Ingrid Richardson), media theory (Mark. B. Hansen, Anna Munster, N. Katherine Hayles), body studies (Elizabeth Grosz), feminist new materialisms (Vicki Kirby, Karen Barad), among others, have all contributed to this conversation. While differing in emphasis, they broadly converge on the idea that the body and the feeling of embodiment are not fixed, enclosed objects, but flexible, open, and responsive processes entangled with their environment.

---

affectively and materially. It is this ongoing negotiation rather than absorption that structures the experience.

One of the most influential empirical demonstrations of embodiment flexibility is the *rubber hand illusion*,[3] in which participants begin to feel ownership over a rubber hand when their real hand and a visible rubber one are stroked in synchrony. This experiment reveals how *proprioceptive coherence* – the felt unity of the body – can be manipulated, demonstrating that the body's somatic boundaries are susceptible to realignments through audiovisual and tactile stimuli. Proprioception (from Latin *propius*, »one's own«, »individual«, and *capio, cepi* – »to receive«, »to perceive«) is one of the sixth senses that is taken for granted the most since a particular organ does not represent it but the body as a whole. It is proprioception that is responsible for the body's feeling of its position and movement in space. Thus, the rubber hand is not a visual illusion but relies on the body's capacity to ›imagine‹ its potential positions and align towards them.

Similar experiments using VR technologies have replicated and expanded these findings,[4] showing how users come to feel that digital avatars, limbs, or full-body representations are their own. These experiments suggest that VR technologies produce (often lingering) transformations in the participant's *somatic self-perception* and *body schema*,[5]that is, the dynamic internal model of the body used to guide movement and interaction. These effects are especially significant because they counteract the initially device-related hermeticism of VR, which can make the user feel disembodied or artificially detached. Through sensory feedback loops, these systems progressively align the user's proprioceptive and affective experiences with the simulation.

3 Matthew Botvinick/Jonathan Cohen (1998): »Rubber Hands ›Feel‹ Touch That Eyes See«, in: Nature 391/6669, p. 756.

4 Such as Julia Reinhard/Dominik Reinhardt/Mel Slater (2020): »Proprioceptive Drift and Embodiment«, in: Frontiers in Psychology 11, n.p.; Tabitha C. Peck/Sofia Seinfeld/Selene Aglioti/Mel Slater (2013): »Putting Yourself in the Skin of a Black Avatar Reduces Implicit Racial Bias«, in: Consciousness and Cognition 22/3, pp. 779–787; Domna Banakou/Parasuram D. Hanumanthu/Mel Slater (2013): »Illusory Ownership of a Virtual Child Body Causes Overestimation of Object Sizes and Implicit Attitude Changes«, in: Proceedings of the National Academy of Sciences 110/31, pp. 12846–12851.

5 For more information about this concept and its differences with body-image consult the text Shaun Gallagher (1986): »Body Image and Body Schema: A Conceptual Clarification«, in: The Journal of Mind and Behavior 7/4, pp. 541–554.

As Ingrid Richardson[6] observes in her work on mobile technologies, such experiences with technology articulate and transform what she calls the *techno-soma*: a hybrid configuration of the sensing body, composed not only of organic matter but also of technological extensions and media environments.

These experimental and theoretical insights have been extended in the context of interactive media art and XR technologies by scholars such as Mark B. Hansen[7] and in our own work.[8] Rather than framing bodily reconfigurations as the substitution of old boundaries with new ones, this line of research invites us to approach the interface not merely as a point of connection, but as a relational condition, a structuring force that modulates how we sense, respond, and become in relation to XR media environments. The focus shifts not only to what the body becomes with these systems, but also how that becoming is orchestrated through the interplay of hybrid materialities and codifications. In this context, the concepts of *interface* and *metastability* provide a framework for analysing the affective and proprioceptive dynamics of user-system interaction.

For one of us, for Ksenia Fedorova, an interface – or interfacing – is a relational condition, not exclusive to digital systems, in which the body and the environment become entangled in a field of dynamic, processual, and affectively encoded exchanges. Drawing from thermodynamics and cybernetics, she argues that interfaces are not stable zones of interaction but »the process of entering into a dynamic framework of relations organised in a procedural structure.«[9] Echoing Johanna Drucker,[10] who argues that interfaces are not access points but events of subjectivity formation, Fedorova emphasises that interfaces not only connect but also transform and structure experiences; they are, in this sense, tactical. This procedural-relational framing shifts our attention away from interfaces as tools for representation or interaction to bodily fields – both affective and proprioceptive – that emerge through ongoing negotiation

6 Ingrid Richardson (2010): »The Hybrid Ontology of Mobile Gaming«, in: Convergence 16/4, pp. 419–430.

7 Mark B. Hansen (2004): New Philosophy for New Media, Cambridge, MA: MIT Press; Id.: Bodies in Code.

8 Ksenia Fedorova (2020): Tactics of Interfacing: Encoding Affect in Art and Technology, Cambridge, MA: MIT Press.

9 Ibid.

10 Johanna Drucker (2011): »Humanities Approaches to Interface Theory«, in: Culture Machine 12, n.p.

between the user and their media-created environment. This dynamic relationality finds a particularly rich expression in XR systems, where unstable yet patterned feedback conditions continually modulate the body's orientation.

But how does this relational and procedural understanding of interfacing actually manifest in the body's experience of XR? What kind of condition enables the body to remain open and in constant adaptation to its hybrid environment? Fedorova addresses this question through the term *metastability*, which she adapts from Gilbert Simondon.[11] She defines metastability as a dynamic equilibrium, one sustained not by isolated components but by the shifting relations among them. Unlike homeostasis, which implies self-regulation toward stability, metastability names a condition of oscillation, readiness to shift, and the system's potential for transformation through relational intensities. Scholars such as Erin Manning[12] have similarly drawn on this concept to explore bodily flexibility, emphasising that bodies are never fully enclosed or stable, but always in flux, always in a constant state of becoming through movement and relation. Fedorova positions metastability specifically within the process of digital interfacing, where perceptual and affective thresholds are continually recalibrated. As she writes, metastability is a »relational scenario for living through that state of confused boundaries and in constant (suspenseful) readiness to respond to changing conditions.«[13] In this light, *metastable interfacing* offers a powerful conceptual tool to describe how users in XR environments do not simply »access« a system but become with it, undergoing proprioceptive and affective transformations that are always in motion.

For Fedorova and Hansen, the embodied flexibility enabled, demanded or triggered by these metastable experiences is materialised through the user's proprioceptive capacities. Proprioception is not a self-contained system of internal feedback. Rather, it is relational and porous, shaped by external cues that are felt as internal. This understanding resonates strongly with the rubber hand illusion experiment, which manipulates the boundaries of *bodily coherence* –what users believe their body is and what it can feel like – by aligning audiovisual and tactile cues.

11 Gilbert Simondon (1992): »The Genesis of the Individual«, in: Jonathan Crary/Sanford Kwinter (eds.), Incorporations, New York: Zone Books, pp. 297–319.

12 Erin Manning (2009): Relationscapes: Movement, Art, Philosophy, Cambridge, MA: MIT Press.

13 K. Fedorova: Tactics of Interfacing, p. 23.

A key aspect of this approach to proprioception is that it is *pre-* or *non-subjective*: it emerges not from a fixed *body-image*, but from a blurring of internal and external orientations of bodily self. This dynamic does not imply a total loss of bodily coherence. Instead, a sense of ownership persists, though not only as active control, but as an affective recomposition of how and where the body feels. In other words, a virtual avatar or limb may be felt as part of one's body not simply because it is controlled, but because it produces proprioceptive effects that transform the user's body schema. The body appears to absorb and expand into, or enfold with, the digital materiality.

In social VR (SVR) platforms such as *VRChat*, metastability manifests not only in proprioceptive experiences but also in existential and performative transformations. SVR platforms give people the opportunity to interact anonymously, primarily via avatars and text (although voice is also possible); they share experiences in the virtual environment, collectively building it or simply exploring and informally communicating with one another. As David Berliner and Maria Erofeeva[14] argue, users undergo shifts in how they perceive themselves and how they perform their identity, shifts that are intimately linked to proprioceptive alterations. Their ethnographic research reveals that avatars in *VRChat* are not merely symbolic representations, but become generative figures through which new or ›truer‹ versions of oneself can emerge.

Participants often describe developing deep attachments to avatars that initially felt artificial. Over time, however, these avatars become like what some describe as a »second skin« or a mask, producing a sense of self-coherence that is not grounded solely in an experience of realism, but in affective familiarity, embodied consistency and identity authenticity. Importantly, this is not a passive transformation but an active, iterative process: many users design and test multiple avatars until they find one that resonates with who they feel they are or aspire to become. In this sense, metastability emerges not only as a state but as a technique – a practised, affectively charged and controlled negotiation through which users craft a coherent sense of self across material and symbolic registers. One participant, for instance, recounts how their long-term use of a cat-like avatar with femboy aesthetics began to merge with their psyche, bodily feelings and performative gestures. After extensive use, they report feeling the two (the avatar and themselves) had become one and the same, stating: »It's

---

14 David Berliner/Maria Erofeeva (2025): »Becoming Otter: Avatars and the Crafting of the Self in Social Virtual Reality«, in: Journal of Material Culture 30/4, pp. 414–438, https://doi.org/10.1177/13591835251377583.

like wearing a mask until it becomes your real face.«[15] Such experiences suggest that VR avatars do not simply extend the body; they restructure the conditions under which the self becomes experientially coherent.

Another aspect of Berliner and Erofeeva's research is how embodied transformation in *VRChat* can manifest as a quest for a »truer« or more »authentic« bodily identity. Several participants report feeling more themselves while »wearing« their avatars than without them. These reports suggest that identity is not only transformed but, in some cases, clarified or achieved through VR. For example, a transwoman whose gender journey predates her use of social VR emphasises how the platform has helped her alleviate her body dysmorphia: »Because in VR, your body is whatever you want it to be... I can look down and see an idealised version of me [...] And it's not just for trans people. If you don't like something about your body, when you're in VR, it's not real. That body isn't there.«[16]

At the same time, other accounts suggest that behaviours and proprioceptive habits cultivated in VR bleed into offline life. One user, for instance, describes unconsciously reaching out to touch a friend's nose at school – a gesture native to *VRChat*'s social etiquette – and feeling instant embarrassment. Such instances resonate with previous VR studies, which have shown that virtual embodiment can alter users' behaviour and proprioception even after the experience ends.[17] These accounts reveal two distinct yet entangled phenomena: one of simulated self-realisation and another of performative and proprioceptive crossover. Both illustrate how metastable interfacing not only expands the user's sense of embodiment but also generates coherence across realities, blurring the lines between simulation and lived identity.

Therefore, we argue that these experiences are not simply the result of immersion; they emerge from a metastable interplay between bodily affect, memory, performativity, technique, and interfacing – a dynamic in which presence and coherence are not given but rather outcomes of repetition and attunement. The user does not become who they »truly are,« but instead who they can become in relation to the VR system's dynamic feedback and affordances.

15 Ibid., p. 20.

16 Ibid.

17 Such as D. Banakou/P. D. Hanumanthu/M. Slater: Illusory Ownership; T. C. Peck/S. Seinfeld/S. Aglioti/M. Slater: Putting Yourself in the Skin of a Black Avatar; J. Reinhard/D. Reinhardt/M. Slater: Proprioceptive Drift and Embodiment.

This process is shaped by ongoing proprioceptive recalibration and performative negotiation. What Berliner and Erofeeva describe is not the revelation of a stable identity, but a practice of self-crafting, in which affective attunements and sensorimotor learning converge to produce a subject who oscillates between feeling momentarily coherent yet never fully settled.

This brings us back to the question of what XR interfaces transform, and how they do so. When read alongside our own work and that of Berliner and Erofeeva, a generative tension emerges: between metastability, as a continual becoming, and simulation, as a staging ground for stable forms of self-expression. XR, in this sense, does not resolve identity; it negotiates it. The simulated body is not merely provisional or illusory; it is a site of potential subjectivity, where the suspension of self through proprioceptive reorganisation coexists with the performance of coherence through *affective orientation*.

2. Simulation as Regulatory Attunement for Embodied Coherence

How does coherence emerge in environments designed through disorientation? In XR systems, users are immersed in spaces where sensory cues, spatial relations and bodily expectations are unstable, fragmented or fluctuating. Rather than being continuous or predictable, experiences in these environments unfold with *metastable interfacing dynamics*. Yet, despite these conditions, users routinely report sensations of wholeness, coherence, and even bodily integration. This paradox invites a deeper question: how does the user come to feel aligned within experiences that actively disrupt sensory continuity? What enables a momentary sense of orientation amidst ongoing instability?

To address this, we turn to the concept of *simulation*. Traditionally, simulation is defined as the creation of a system or model that imitates or predicts the behaviour of real-world processes – often used in scientific and technological domains such as climate forecasting or aeronautics.[18] In media theory,

18 Stephan Hartmann (1996): »The World as a Process: Simulations in the Natural and Social Sciences«, in: Rainer Hegselmann/Ulrich Mueller/Klaus G. Troitzsch (eds.), Modelling and Simulation in the Social Sciences from the Philosophy of Science Point of View, Dordrecht: Springer, pp. 77–100; Paul Humphreys (2004): Extending Ourselves: Computational Science, Empiricism, and Scientific Method, Oxford: Oxford University Press.

especially following Jean Baudrillard,[19] simulation takes on a more critical dimension: a self-referential, hyperreal image or illusion that replaces reality. In VR theory, as David J. Chalmers[20] suggests, simulation fuses both these logics. Virtual realities are not mere illusions, he claims, but genuine realities in themselves – ontologically distinct yet experientially real, built through computational models that afford interaction, change, and subjective presence.

However, we want to shift the emphasis: from simulation as a descriptor for technical capacities to simulation as a body-system relational process. Our concern is not whether simulation mirrors, predicts or constructs a substitute for reality, but how it structures experience through bodily engagement. Simulation becomes meaningful not as a model either to be compared to an original or in its own conditions, but as a *regulatory field*: a dynamic interface in which coherence is felt, not found. It is not a preexisting structure into which users enter, but a relational ecology that emerges through calibration and feedback. Simulation, in this sense, is less a representational technology than an *affective technique*, one that modulates perception, orientation and coherence.

This framing returns us to the questions of coherence in metastable systems, which aligns with broader philosophical accounts of perception and embodiment. Henri Bergson's notion of *qualitative multiplicity*[21] challenges the idea that experience is composed of discrete, stable units. Instead, he describes consciousness as a continuous yet heterogeneous flow: a lived synthesis that holds difference together without collapsing it. Coherence in this framework is not given but composed; it is not the product of cognitive clarity but of intuitive navigation through shifting states. Similarly, Timothy Morton's[22] ecological philosophy stresses the irreducibility and interdependence of systems, emphasising that beings are always entangled in relations they cannot fully perceive or control. These thinkers share the claim that coherence is not a property but a process – a felt resolution that emerges from complexity rather than despite it.

---

19 Cf. Jean Baudrillard (1994): Simulacra and Simulation, Ann Arbor: University of Michigan Press.

20 Cf. David J. Chalmers (2022): Reality+. Virtual Worlds and the Problems of Philosophy, New York: W. W. Norton & Company.

21 Cf. Henri Bergson (1912 [1903]): An Introduction to Metaphysics, New York: G. P. Putnam's Sons.

22 Cf. Timothy Morton (2016): Dark Ecology: For a Logic of Future Coexistence, New York: Columbia University Press.

We approach these dynamics through two interrelated concepts: *attunement*[23] and *orientation*.[24] Attunement, drawing from affect theory and process philosophy, refers to the body's ongoing responsiveness to its environment. So, the capacity to adjust, recalibrate, or synchronise in response to shifting conditions. It is not a passive reception, but a form of micro-negotiation between internal states and external cues. Erin Manning employs the concept of attunement to explain how the body »gets in tune« with the forces that bring about an event.[25] Orientation, meanwhile, refers to the directional force of these negotiations: how bodies are drawn toward or away from specific relational alignments, perceptual anchors, or rhythms. Sara Ahmed refers to how subjects are positioned in space and time by the objects, bodies, and environments they inhabit. Orientation is not only about physical direction, but about the affective and perceptual vectors that shape how we inhabit the world.

Placing the analysis of technology at the intersection of process philosophy, such as Bergson and Manning, and phenomenology, like Ahmed, allows us to discuss the virtual not as an opposite of the real, but as a potential yet to be activated (a thought elaborated extensively by Brian Massumi[26]). Whereas *VRChat* foregrounds the exploration of potential identities, these processes of becoming can also be understood as operating within a dynamic interplay between the actual and the virtual – not as endpoints along a continuum, but as

---

23 Cf. E. Manning: Relationscapes.

24 Cf. Sara Ahmed (2006): Queer Phenomenology: Orientations, Objects, Others, Durham: Duke University Press.

25 We want to thank Annette Urban and Manischa Partowi for pointing us to the discussion of *alignment* in Julia Reich/Manuel van der Veen (2023): »A Kind of Mixed, Intermediate Experience: On the Entanglement of Image and Bodies«, in: Yearbook of Moving Image Studies (Mixed Reality Images), Darmstadt: Büchner, pp. 92–114. Their notion of alignment offers a valuable account of how the viewer and the image-body become performatively synchronised within MR environments – a choreography through which perception and movement are co-constituted across real and virtual domains. While our argument resonates with this relational framing, we extend it through the concept of *attunement*, which foregrounds the affective and ecological dimensions of such coupling. Rather than describing spatial and movement synchronisation, attunement names an ongoing modulation through which the body sustains coherence under metastable and shifting conditions, emphasising that the coherence is not performed but felt.

26 Cf. Brian Massumi (2002): Parables for the Virtual: Movement, Affect, Sensation, Durham, NC: Duke University Press.

interwoven modes of embodiment and expression. Situations in XR environments have a more powerful psychological effect than traditional forms of art, such as literature, theatre, and cinema, which also involve elements of identification with the character and techniques that immerse the audience in the storyworld. In XR, not only the minds but also the very bodies of users are actively involved, contributing their states to the overall experience and its coherence. While participants in *VRChat* attune their avatars, i.e., their potential selves, to what feels »true« and authentic from within, it is also important to note that these gestures and movements are shaped by volumetric tracking systems, animation presets, and control regimes that mould how the body can express itself in virtual experiences. Yet, within these constraints, users still draft rhythms of coherence that feel experientially authentic.

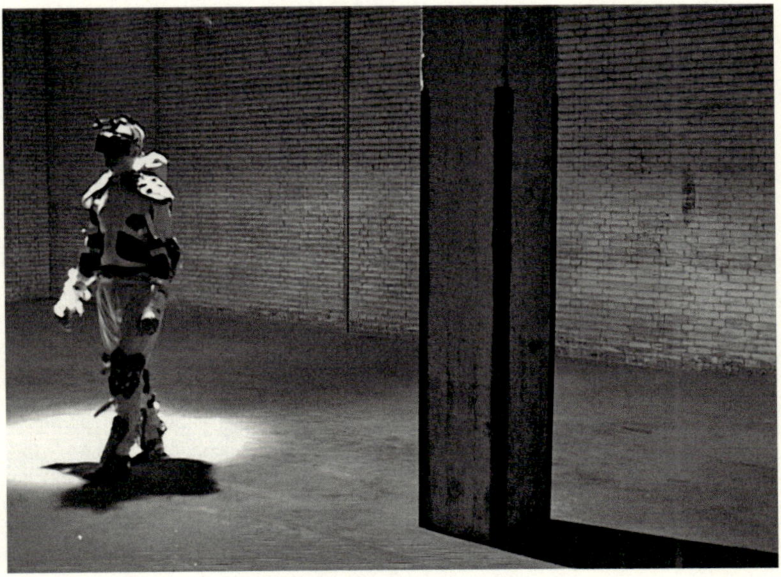

fig. 1: Marnix de Nijs: Exercise in Immersion 4, 2006-2007

Marnix de Nijs's *Exercise in Immersion 4* (2006-2007; hereafter *EI4*) offers a peculiar instantiation of the role of the body in XR and, with that, of simulation as attunement. Described as an »art-game,« the piece invites the users into a hybrid installation where physical movement, responsive spatialisation, and gamified feedback become central. The setup is deceptively simple. Participants wear a custom-designed crash suit and a head-mounted display (HMD), navigating an expansive warehouse space equipped with motion tracking and pre-rendered visual overlays (fig. 1). All technical elements were made by the artist himself, reflecting the media art of the time and allowing greater creative control. Initially, the simulated environment closely resembles the physical warehouse. This 1:1 mapping mimics spatial alignment and fosters an expectation of stability.

But this alignment quickly begins to slip. As the participant moves through space, the visuals in the HMD start to deviate from the warehouse's architectural reality. Abstract, shifting geometries gradually replace familiar structures. Visual landmarks disappear. Gravity fluctuates. The longer the user moves, the more the environment disorients them. The key is not fidelity, but progressive dislocation – a slow, deliberate drift between visual expectation and proprioceptive feedback.

This growing discrepancy does not break the simulation; it intensifies it. The user's challenge is not to remain immersed in a coherent illusion, but to recalibrate and re-orient in real time. Movement becomes both exploratory and tactical, guided not by visual continuity but by the shifting sensorium of cues. It is motivated by visual elements—referred to as ›bionts‹—that float in space and guide users in avoiding collisions with physical structures that are no longer visible in the headset (figs. 2–3). These bionts were semi-autonomous, slime-like agents that the user collected throughout the experience, much like Pac-Man. Once collected, these bionts swarm around the user, acting as perceptual extensions: they alert the player to spatial obstacles, buffer against collisions, and function as indicators of relational proximity. Their presence effectively scaffolds a relational awareness between the user's body and the hybrid space, enhancing orientation even as the environment slips into abstraction. Importantly, the experience is designed to »fail.« When a participant collides with a wall or a pillar, the visual environment is momentarily re-synchronised with the original warehouse. This moment of realignment is not a return to immersion as coherence but an apparent »reset« that demonstrates how simulation works as a process of rhythmic punctuation between consistency and chaos. The bionts not merely guide; they modulate the body's sense of proximity and relational coherence, shaping an ecology of feedback among the user, the system, and the semi-autonomous agent. Furthermore, their presence, as well as the strategy of taking participants out of their comfort zone, is typical of De Nijs' work and speaks to the specificity of this project as an artwork. It is both the creative choices, the playful element, and the counterintuitive motivation approach that make this work an open-ended experiment, enabling unique experiences and new observations about the capacities of the body-mind to modulate and re-align.

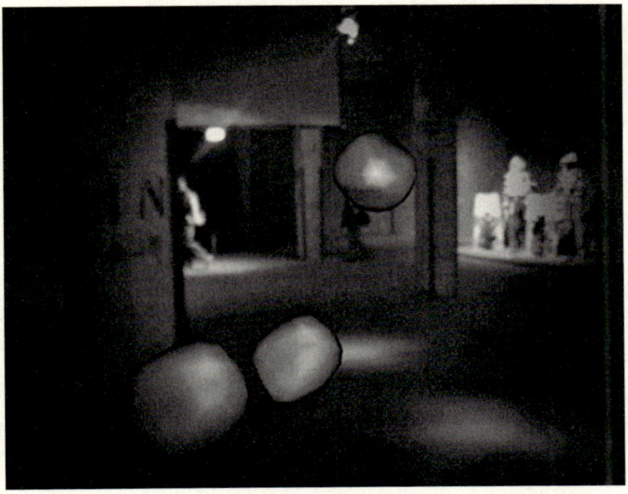

fig. 2: Marnix de Nijs: Exercise in Immersion 4, 2006-2007

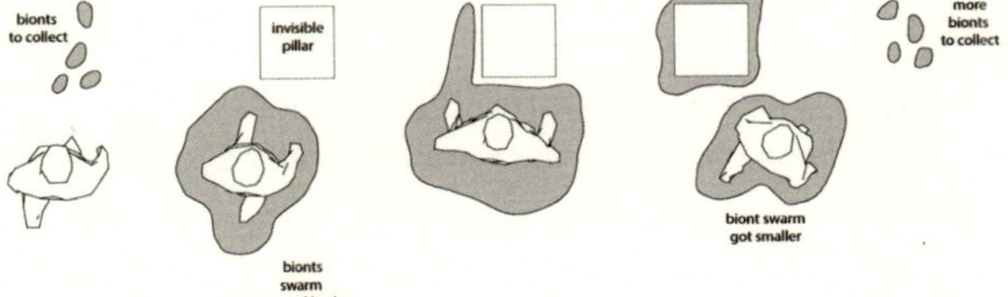

fig. 3: Marnix de Nijs: Exercise in Immersion 4, 2006-2007

Simulation, from the Latin *simulatio*, meaning ›imitation‹, ›pretence‹, or ›representation‹, and *simulare*, ›to make like‹, has long been understood through two dominant traditions. The first treats simulation as representation – a visual or perceptual imitation of the real. In this tradition, most commonly associated with Baudrillard, simulation – unlike *mimesis*, which works as a copy of reality – is the creation of a seamless illusion that conceals the absence of reality behind its image. The second, more technical tradition defined simulation as modelling – the construction of systems (physical, computational, or conceptual) that mimic or predict the behaviour of real-world processes.[27] David J. Chalmers notes that simulations do not necessarily replicate an existing process, but they can be used to create novel processes in their own right.[28] From engineering to medicine to art, simulation enables controlled experimentation by abstracting and formalising the dynamics of the target system.

These traditions persist in XR discourse. Many VR applications aim to replicate reality with high fidelity, while others use simulation to model possible experiences, environments, and identities. Chalmers, in *Reality+*, engages both. He suggests that virtual environments are not merely illusions or models; they are real realities in their own right – ontologically valid spaces that support agency, interaction, and experience. For Chalmers, simulations are real because they instantiate systems with consistent causal dynamics, regardless of whether those dynamics resemble the physical world.

We do not reject this ontological framing. However, our focus lies elsewhere: not on what simulation is in its technical capacities, but how it is lived. We shift the attention from simulation as a technological system to simulation as an embodied relation. That is, simulation becomes meaningful not because it represents or predicts a world, but because it modulates orientation; it structures how bodies move, feel, and become in dynamic relation to fluctuating environments.

Simulation, in this view, is not a model into which the user steps and manipulates to get coherent results. It is a *regulatory field*: a dynamic interface that is co-constituted through action and sensation. It does not precede experience as a fully formed structure; it is felt and enacted through metastable interfacing – a process of continuous calibration between the body and the system. In this sense, simulation is not only what appears in the headset or in the coding

27 Cf. S. Hartmann: The World as a Process.

28 Cf. D. J. Chalmers: Reality+.

processes; it is what emerges in the rhythm of interaction. Coherence is not found but practised.

This ongoing calibration is best understood through Gilles Deleuze's notion of *modulation*,[29] which shifts the emphasis from disciplinary control to continuous affective adjustment. Applied to VR experiences, modulation is not the total elimination of control or disciplinary mechanisms by technological affordances, but rather their reconfiguration as a dynamic, responsive force – one that adapts, tunes, and reshapes how bodies interact with systems. In simulation, modulation makes coherence possible: not through the imposition of structure, but through its relational ecology, which attunes the user's sensorium through shifting gradients of affect and orientation. Ahmed's concept of *orientation*[30] furthers this view, suggesting that bodies are not merely in space, but are continually shaped by what they are drawn toward or repelled from. Manning's work on proprioception extends this further: movement and sensation do not originate solely in the body, but co-emerge with the environments that modulate them.[31] Simulation as attunement, then, is not passive reception of cues, but a field of modulation that enables coherence to be enacted – not once, but again and again, across changing conditions.

The case of *EI4* complicates standard accounts of immersive VR, which often prioritise »seamless immersion« and perceptual continuity. In other words, VR experiences tend to avoid a coherence mismatch between what the user sees and feels in space, thereby avoiding potential physical collisions. This phenomenon is a familiar experience for habitual VR users who adapt their living rooms for VR play or use. However, for *EI4*, these collisions do not »break immersion,« they are not treated as interruptions but as integral parts of the simulation. We suggest that, besides complicating the boundaries between »real« and »virtual« materialities and ontologies, it also materialises that coherence in XR is not given by realism or consistency, but rather by attunement capacities to new patterns and rhythms. So, attuning to the experience as a simulation here works as a regulatory force that orients the body towards readjusting to new logics, thereby existing in coherence within these constantly changing, metastable relations.

Collision in *EI4* exemplifies simulation as attunement, as a way that structures the ways users feel coherence. Rather than presenting the XR experience

29 Gilles Deleuze (1992): »Postscript on the Societies of Control«, in: October 59, pp. 3–7.

30 Cf. S. Ahmed: Queer Phenomenology.

31 Cf. E. Manning: Relationscapes.

as a seamless world, *EI4* operated through a cycle of rupture and orientation. The user does not become part of a stable or predictable reality – at least not at the beginning. They practice coherence by continuously attuning to the simulation, not only to the »new« logics of digitality but also to their ambivalence and their reformulation of »preexisting« ones. Coherence here is not a property of the environment, but an emergent feeling that the changing forces of VR are transforming the world the user metastably becomes with.

In XR environments, we argue, simulation is modulation. It not only presents an immersive image of the world, but it also orchestrates the conditions of attunement. Even when environments are fragmented, unstable, or unfamiliar, the simulation guides the user toward momentary coherence through the regulation of cues and perceptual rhythms. Coherence in this sense is not the goal of simulation, but its effect – a felt outcome that emerges through patterned instability. So, if simulation is not a representational layer but a regulatory ecological force, then its central operation lies in how it orients the bodily feelings. We propose that simulation in XR functions not as a visual or cognitive illusion, but as a system that attunes the user toward momentary coherence.

This is what we mean by simulation as attunement, not the reproduction of a stable environment, but the dynamic calibration of the user's sensorium to patterns of instability. Simulation creates rhythms, thresholds and resistances, not to resolve instabilities, but to allow the body to learn how to navigate them. The user is not attuning to a fully realised world, but rather attuning into the simulation itself – a field of forces, where coherence emerges through adjustments.

This view shifts the emphasis from fidelity to feedback. Simulation, as we understand it here, operates by modulating tensions – between visual and proprioceptive perception, between prediction and response, and between coherence and disorientation. The role of the user is not to detect whether the world is »realistic,« but to develop rhythms of action within it. The simulation, in turn, becomes a kind of training field for these rhythms. It is a space where the user's orientation is always slightly off-balance, but never outside the possibility of temporary alignment.

Importantly, attunement is not reducible to interactivity or immersion. It is not a feature of the software, but a relational achievement – a process that unfolds over time between the user and the MR environment. Affective attunement does not guarantee clarity or stability; instead, it enables coherence to be felt amidst difference.

The logic of simulation as attunement in *EI4* resonates with the social and identity-oriented practices of users in *VRChat*. As Berliner and Erofeeva[32] show, users on social VR platforms often describe a transformation of their sense of self. We suggest that this attunement does not solely emerge from inhabiting an alternate world, but rather from learning to perform in coherence. Their ethnographic research demonstrates how avatars are not merely visual representations, but sites of sensory alteration and learning. Over time, avatars become what one participant describes as a »second skin,« a phenomenon that emerges not solely through immersion but through repeated, embodied negotiation of form, posture, movement, and interaction.

This dynamic mirrors the logic of *EI4* in a different modality. Where *EI4* »trains« the user to recalibrate spatial and proprioceptive orientation under destabilised conditions, *VRChat* users recalibrate affective and performative orientations within unstable and unfamiliar avatar bodies. In both cases, simulation is not a container, but a structuring becoming, one that, through attunement with it, allows coherence to emerge as a felt structure.

The metastable oscillations between logics and materialities are not only processes of coherence and breakdowns; these movements are compositional elements oriented by the affective forces of simulation. In other words, the simulation's »success« lies in its ability to guide the user through misalignments, not in its capacity to hide them. So, XR as simulation becomes an affective infrastructure through which bodies learn to feel situated, even under conditions of constant transformation.

3. Haptics in VR

To specify the effects of embodiment in VR, it is helpful to focus on a sense that foregrounds the body's physicality: the haptic sense. It is through haptic sensations evoked by movement, gesture, and touch, or through skin sensations, that the body manifests its presence. But more than any other modality, haptics grounds the body not only in space, but in relational orientation. If, as we have argued, simulation in XR operates less as representational fidelity and more as attunement, then haptics becomes a critical site where that attunement is enacted, felt and regulated. This raises important questions: Is it possible to adequately transmit a haptic sensation and the associated sense of

32 Cf. D. Berliner/M. Erofeeva: Becoming Otter.

intimacy? What does the simulation of a haptic sensation do to the sense of the self and the borders of the self? In what follows, we aim to explore how touch, gesture, and movement function not only as immersive techniques but also as relational practices of bodily orientation within metastable environments.

The rubber hand illusion, De Nijs' work and some of the experiences in *VR-Chat* all have haptic sensation in their core. One of the most elusive senses, the haptic sense (from Greek *haptós*, meaning palpable) encompasses phenomena of touch and tactility, as well as other somatic sensations that extend beyond the body's surface, such as kinaesthesia, the sense of balance, and proprioception. More than any other, the haptic sense is grounded in the body as a whole. It is impossible to feel space without first being aware of the starting point for that orientation – one's own body. Although it seems that one is primarily exposed to the virtual space through vision, the haptic sense immediately comes into play as the beholder or their avatar begins to move or is touched.

As proposed above, we see simulation not as a replication of sensations of touch and movement in an imagined reality to be attuned to, but rather as a means of attuning to the sensations of touch and movement themselves. The very technique of simulation regulates bodily engagement. It organises the emergence of proprioception toward fleeting but felt coherence, generating a relational field through which touch-like self-orientation becomes possible. The effect becomes ever more salient if the body is activated in a direct physical sense, not only through vision, i.e., via the neurological connections within the brain that trigger sensations of movement or touch felt purely in the brain, rather than on the skin or within muscles.

While the above-discussed cases involve elements of haptics, it plays an even more prominent role in projects driven by dance and choreography. Such projects typically feature a pre-recorded dancer who guides the user through the space and interacts with them, for example, by looking directly into the user's eyes, stretching their arms to reach out to the user's hands, gesturing an invitation to follow, or simply allowing them to be observed closely. Below are examples of just two creative projects in which an audience member is involved in dance, either through simple gestures that explore the pre-recorded dance freely or by moving together with both on- and off-screen partners.

Dust (2016–17) by Andrej Boleslavský and Mária Júdová is a 4-minute immersive experience that places the user in a space filled with tiny floating particles (fig. 4). It is possible to see the real-time simulations of one's arms and legs as well as another human figure, who explores this peculiar space through dance, as if inviting the user to do the same. Volumetric recordings of dancers and photogrammetric imagery of the building create the aesthetic effect of a porous space, where there are no clear borders between objects, and the contours appear to flicker. The metaphor of dust here stands for the invisible particles that make up our planet, the »stardust« that once became Earth. According to the authors, the audience was invited to take the perspective of such a particle and to come extremely close to the dancers, to watch them from unusual angles (not possible in real life), and even to step ›inside‹ the dancers' bodies. The permeability of bodies in this virtual environment prompts one to recalibrate and reattune the borders of one's own body through active gestures and movement. The vibrating space and oscillating shapes epitomise here, quite literally, the sense of metastability. While the audience member can still feel solid ground under their feet, the visual field welcomes movements that challenge the solidity of matter.

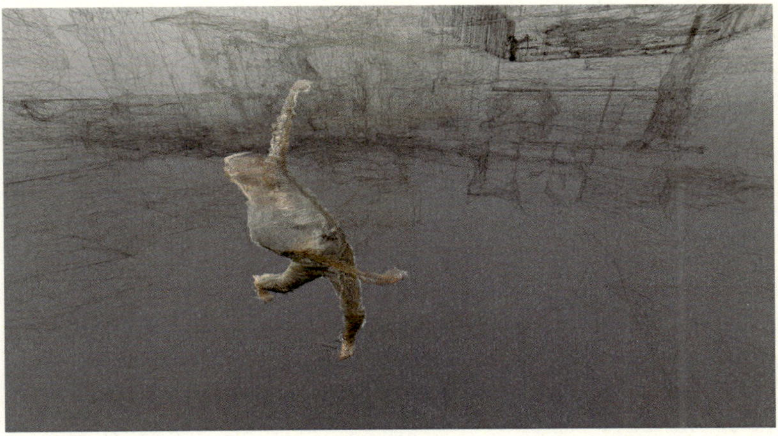

fig. 4: Andrej Boleslavský and Mária Júdová: Dust, 2016

Another example of a VR experience involving dance is *Eve 3.0* (2024) by Margherita Bergamo Meneghini (fig. 5).[33] Instead of only witnessing dance

33 We thank Tingyu Li for bringing Margherita Bergamo's work to our attention, as well as Berghamo herself for presenting this project in Leiden in December 2024.

within the VR, the audience is encouraged to participate in actual dance, synchronising the movements of the figures within the headset with the movement of one's own body. The experience begins with you, the audience member, sitting in a chair or on a bench in a park within the VR story. There is a group of teenagers dancing under a big tree, and at one point, one girl comes to you to pick up her forgotten notebook from your hands. As it happens, you feel someone's hands take yours and invite you to stand up and move with them. This episode is part of a more complex story that involves teenagers stealing the diary and teasing its author. The viewer must take a stance in this social dynamic. It is thus a hybrid performance that happens both in VR and in real life. The authors describe it as a collective dance that is designed to make one »reflect on common and widespread extreme states of consciousness such as addiction, anxiety, depression, obsession, jealousy and paranoia.«[34] These affects are not depicted precisely but rather through an abstract language of dance. This form of allusion to the potential of VR to let us live through various behavioural scenarios is worth noting: the story allows us to recognise particular psychological states at the cognitive level, but the choreography immerses the very body in at least one of them at a time, letting one experience them intuitively and on a more subtle subconscious level.

What interests us in this work is the role of the haptic component in the blending of realities and the psychological effects that this conversion may entail. Active bodily participation in the story necessitates tuning in to the character whose position you are taking – someone to help return the intimate records – the diary – to a safe environment in a joint dance. You thus have to find coherence between your physical body, your usual habitual self and the role you are temporarily given to play. This happens here through alignment of simulation – as you see your own arms in the space of the story, filled with other, virtual characters, – and direct sensation of touch. Thus, touch, as in the rubber hand illusion, once again calls for proprioceptive alignment and the matching of the actual and potential realities.

34 Margherita Bergamo Meneghini (2024): »Eve 3.0«, in: compagnievoix.com. Online: https://compagnievoix.com/en/projects/creation/eve-3 (last access: 13.10.2025).

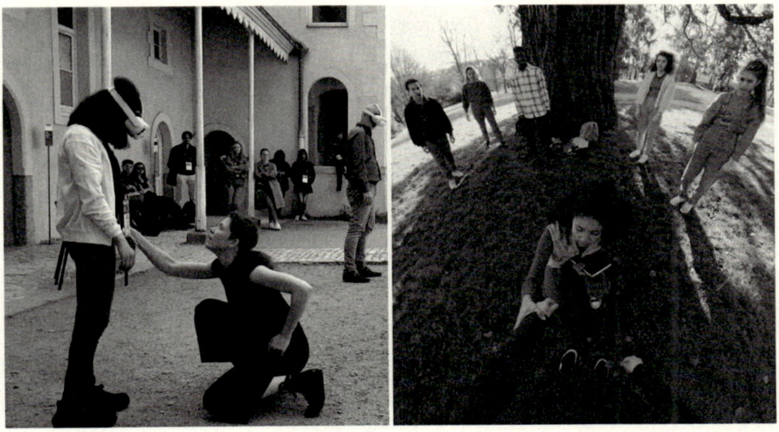

fig. 5: Margherita Bergamo Meneghini, Veronica Boniotti, John Desnoyers-Stewart, and Daniel González-Franco: Eve 3.0, 2023

In media studies, the simulation of haptics and touch has been continuously seen as one of the next frontiers.[35] Unlike vision or hearing, haptics operates on a deeper, more subtle physical level. An overview of the issues and already available hardware is beyond the scope of this text. Instead, the aforementioned projects aim to draw attention to the distinction between imagined and real sensations of touch, as well as to the role of movement and gesture within the same haptic continuum, yet without direct skin sensation. All these haptic elements require bodily activity, necessitating awareness of one's presence and active engagement.

In the history of philosophy, from Aristotle to Merleau-Ponty, touch has been associated with immediacy and indivisibility, as it enables one to feel the connection with the world, which is as close as the surface of one's body. It is, however, worth noting an aspect brought up by Jean-Luc Nancy: that touch also implies separation – an »originary spacing,« *partage*, or »division,« »sharing out« between the two touching parties.[36] Touch is not only contact but also the

35 Cf. David Parisi/Mark Paterson/Jason Archer (2017): Haptic Media Studies, London: Routledge.

36 Jacques Derrida (2005): On Touching: Jean-Luc Nancy, Stanford: Stanford University Press, p. 156.

»in-between« that enables it. Through touch and the haptic sense more generally, we can modulate our connectedness to the world and the ways it impacts us.

In *Dust* and *Eve 3.0*, touch is both simulated and physical. It is reasonable to assume that the authors deliberately sought to highlight this tension between the two, emphasising the significance of the irreproducible inner experience of direct physical sensations. The tension between realities resonates with the conflict between connectivity and distance in touch, as highlighted by Nancy.

Just as in the case of the other examples explored in the earlier sections, simulation in *Dust* and *Eve 3.0* asks us to find oneself anew not only in the reality of the storyworld, but in the physical reality of one's own body and one's own self. Going through the pre-choreographed and improvised movements, you literally perform this coherence, integrating, in a metastable condition, the elements of not only one imagined and physical self but also its multiple *potential* versions.

4. Conflict and Tension: Discovery vs. Becoming

This understanding of simulation as attunement allows us to revisit the tension at the heart of XR embodiment: the pull between metastability, as a continual becoming, and simulation as a site for apparent stability and self-expression. If metastability describes the body's readiness to transform – its openness to new orientations and relational compositions – simulation offers, paradoxically, a scaffold for coherence. Even as it destabilises sensory input or alters spatial logic, simulation provides sufficient rhythmic, affective, and perceptual consistency to allow a subject-user to emerge. In *VRChat*, this might take the form of stable identity performances through avatar use; in *EI4*, it's the body learning to find proprioceptive rhythm within an environment that continually shifts.

Crucially, these are not mutually exclusive processes. The user becomes coherent not by escaping instability, but by learning to inhabit it – by attuning to simulation's modulated cues and treating them as provisional grounds for action. Simulation, then, is not the opposite of becoming – it is what allows becoming to be *felt* as coherent. The user does not become coherent by overcoming instability, but by tactically navigating it. And it is precisely this lived contradiction that defines the experiential power of VR today.

Returning to Jean-Luc Nancy's idea of touch helps illuminate this paradox. For Nancy, touch always implies partition or spacing, a shared distance that both connects and separates. Touch is not fusion, but partition; not collapse but a relational tension held across difference. Touch is not a chaotic combination; it presumes that the touching elements remain distinct, that a gap persists even in contact. In this light, coherence in XR – the »harmonious« touch between materially different elements – is constantly haunted by distance. The body senses and becomes with both contact and separation.

This insight reframes our tension. Simulation may scaffold a felt coherence, but it cannot erase the spacing between body and code, subject and avatar, expectations and sensation. That gap – that collision – is not a flaw; it is what materialises coherence. It demonstrates that coherence is not a given, but a practice – something to be modulated, felt, and continuously re-achieved. So, simulation is not a representation; it is an attunement. Instability is not noise, but the condition through which simulation orients the user. In other words, it is what demonstrates that coherence is a technique, something to be practised rather than assumed. Simulation does not erase differences; it re-orients through them.

Thus, the tension between metastability and coherent performance is not structured as a binary opposition. These are not endpoints along a spectrum from chaos to order. They are coextensive dynamics in the process of *embodying with* XR technologies. It is a process that is neither fully chaotic nor metastable, nor the representation of preexisting structures. The system and the body are not in a completely unstructured fusion, nor do they remain as completely discrete and pre-structured elements. XR does not guarantee wholeness; it modulates the terms through which coherence is possible. So, this tension is not to be resolved, but it is integral to the felt coherence. XR does not deliver self without gaps; it offers a site for the »spacing of coherence« – a coherence that is always already partial, always touched by difference, always attentive to the spacing that makes the relation possible.

The cases we have examined make this tension palpable. For instance, in *Exercise in Immersion 4*, the user's body navigates a simulation built on rupture – collisions, resets, and drifting alignments – where coherence is achieved not through fidelity but through rhythmic recalibration. In *VRChat*, users engage in parallel practices of identity attunement: crafting avatars, gestures, and social expressions until a »second skin« begins to feel authentic, even as its limits remain permeable. In both instances, the biological and the digital never fully fuse, though they may *feel* aligned momentarily. It is this irreducible spacing –

Nancy's original distance – that underpins the possibility of coherence. Similarly, in *Dust* and *Eve 3*, active body movement helps to increase the effect of being close and yet remaining apart. Although moments of tactile and gestural alignment blur the boundary between the virtual and the physical – especially in *Eve 3*, when another person physically holds the user's hands – there remains a persistent sense of separation, a lingering »gap« that materialises disbelief or unreality, even within moments of intimacy. This way, simulation does not erase difference but stages it, offering a patterned field where bodies and selves practice alignment within instability. XR does not deliver a seamless self; it provides a site of coherence that is always partial, always touched by difference, and made livable through attunement.

Conclusion: Simulation as Practice

Reframing simulation as attunement invites a shift in how we conceptualise XR; not as a domain striving for realism, but as a site of affective, bodily practice. Coherence, as we have argued, is not the outcome of seamless representation, but the product of a modulation: a rhythm the body learns to inhabit. The simulation is not just what appears in the headset, but the patterned conditions that guide bodily orientation and self-making. One literally forges one's identity, as in *VRChat*, or discovers one's potential as a participant of a collective dance, as in *Eve 3*.

This perspective holds practical stakes. For artists and XR designers, it suggests that immersion need not hinge on fidelity or continuity. Disorientation, rhythmic breakage, and sensorial misalignment are mattering tools; not disruptions to overcome but textures to work with. Designing for attunement rather than illusion foregrounds the body's role in the construction of experience.

For media theorists, curators, and educators, this reframing also reshapes evaluative criteria. The critical question is no longer how accurately XR mirrors »reality«, but how it scaffolds new modes of relationality. Simulation, in this view, becomes a technique of perception, a way of composing the body across affective, proprioceptive, and technical fields of difference.

To treat simulation as attunement is not only to describe what XR does, but to ask how it might feel otherwise, and what new rhythms of coherence we might learn to inhabit.

Bibliography

Ahmed, Sara (2006): Queer Phenomenology: Orientations, Objects, Others, Durham: Duke University Press.

Banakou, Domna/Parasuram D. Hanumanthu/Mel Slater (2013): »Illusory Ownership of a Virtual Child Body Causes Overestimation of Object Sizes and Implicit Attitude Changes«, in: Proceedings of the National Academy of Sciences 110/31, pp. 12846–12851. DOI: https://doi.org/10.1073/pnas.1306779110

Baudrillard, Jean (1994): Simulacra and Simulation, Ann Arbor: University of Michigan Press.

Bergson, Henri (1912 [1903]): An Introduction to Metaphysics, New York: G. P. Putnam's Sons.

Berliner, David/Maria Erofeeva (2025): »Becoming Otter: Avatars and the Crafting of the Self in Social Virtual Reality«, in: Journal of Material Culture 30/4, pp. 414–438. DOI: https://doi.org/10.1177/13591835251377583

Chalmers, David J. (2022): Reality+. Virtual Worlds and the Problems of Philosophy, New York: W. W. Norton & Company.

Deleuze, Gilles (1992): »Postscript on the Societies of Control«, in: October 59, pp. 3–7.

Derrida, Jacques (2005): On Touching: Jean-Luc Nancy, Stanford: Stanford University Press.

Drucker, Johanna (2011): »Humanities Approaches to Interface Theory«, in: Culture Machine 12, n.p.

Fedorova, Ksenia (2020): Tactics of Interfacing: Encoding Affect in Art and Technology, Cambridge, MA: The MIT Press.

Gallagher, Shaun (2005): How the Body Shapes the Mind, Oxford: Oxford University Press.

Hansen, Mark B. (2004): New Philosophy for New Media, Cambridge, MA: MIT Press.

Hansen, Mark B. (2006): Bodies in Code: Interfaces with Digital Media, New York: Routledge.

Hartmann, Stephan (1996): »The World as a Process: Simulations in the Natural and Social Sciences«, in: Rainer Hegselmann/Ulrich Mueller/Klaus G. Troitzsch (eds.), Modelling and Simulation in the Social Sciences from the Philosophy of Science Point of View, Dordrecht: Springer, pp. 77–100.

Humphreys, Paul (2004): Extending Ourselves: Computational Science, Empiricism, and Scientific Method, Oxford: Oxford University Press.

Jarvis, Liam (2021): Immersive Embodiment: Theatres of Mislocalized Sensation, London: Palgrave Macmillan.

Manning, Erin (2009): Relationscapes: Movement, Art, Philosophy, Cambridge, MA: MIT Press.

Massumi, Brian (2002): Parables for the Virtual: Movement, Affect, Sensation, Durham: Duke University Press.

Morton, Timothy (2016): Dark Ecology: For a Logic of Future Coexistence, New York: Columbia University Press.

Nancy, Jean-Luc (2008): Corpus, New York: Fordham University Press.

Peck, Tabitha C./Sofia Seinfeld/Selene Aglioti/Mel Slater (2013): »Putting Yourself in the Skin of a Black Avatar Reduces Implicit Racial Bias«, in: Consciousness and Cognition 22/3, pp. 779–787. DOI: https://doi.org/10.1016/j.concog.2013.04.016

Reich, Julia/Manuel van der Veen (2023): »A Kind of Mixed, Intermediate Experience. On the Entanglement of Image and Bodies«, in: Yearbook of Moving Image Studies (Mixed Reality Images), Darmstadt: Büchner, pp. 92–114.

Reinhard, Julia/Dominik Reinhardt/Mel Slater (2020): »Proprioceptive Drift and Embodiment«, in: Frontiers in Psychology 11, n.p. DOI: https://doi.org/10.3389/fpsyg.2020.00258

Richardson, Ingrid (2011): »The Hybrid Ontology of Mobile Gaming«, in: Convergence 17/4, pp. 419–430.

Simondon, Gilbert (1992): »The Genesis of the Individual«, in: Jonathan Crary/Sanford Kwinter (eds.), Incorporations, New York: Zone Books, pp. 297–319.

Slater, Mel/Sylvia Wilbur (1997): »A Framework for Immersive Virtual Environments (FIVE): Speculations on the Role of Presence in Virtual Environments«, in: Presence: Teleoperators and Virtual Environments 6/6, pp. 603–616.

Slater, Mel/Daniel Perez-Marcos/H. Henrik Ehrsson/Maria V. Sanchez-Vives (2010): »Inducing Illusory Ownership of a Virtual Body«, in: Frontiers in Neuroscience 4, pp. 1–8. DOI: https://doi.org/10.3389/fnins.2010.00010

Image Credits

fig. 1–3: © Marnix de Nijs
fig. 4: © Andrej Boleslavský and Mária Júdová

fig. 5: © Margherita Bergamo Meneghini, Veronica Boniotti, John Desnoyers-Stewart, and Daniel González-Franco, photo: © Compagnie Voix.

Playing VR Against Itself
Immersive Environments between Hypermediation and Remediation

Federica Cavaletti, Pietro Conte, Andrea Pinotti

This study[1] investigates the paradoxical dynamics of empathy, illusion, and self-reflexivity within contemporary virtual reality (VR) storytelling. After outlining the ideological and technological premises that have led many artists and filmmakers to conceive of VR as an ›empathy machine‹, the essay questions this assumption by highlighting the structural contradictions inherent in immersive media. The first section analyses the affective and ideological implications of empathy-oriented VR projects, showing how their apparent immediacy conceals deeply mediated frameworks. The second and third sections then turn to two case studies – Michał Stankiewicz's *Wish You Were Here* and John Hsu and Marco Lococo's *Great Hoax: The Moon Landing* – that deliberately challenge the rhetoric of transparency by exposing, respectively, the illusion of immersion and the instability of authenticity. Through these

1 This article was written in the framework of the research project *AN-ICON. An-Iconology: History, Theory, and Practices of Environmental Images*. The project has received funding from the European Research Council (ERC) under the European Union's Horizon 2020 research and innovation programme (grant agreement No. 834033 AN-ICON). Among the authors, Federica Cavaletti further received funding by the European Union (ERC-2023-POC *TIMELAPSE*, Project number: 101137707). Views and opinions expressed are those of the author(s) only and do not necessarily reflect those of the European Union or the European Research Council Executive Agency. Neither the European Union nor the granting authority can be held responsible for them. Both *AN-ICON* and *TIMELAPSE* are hosted by the University of Milan, Department of Philosophy *Piero Martinetti* (Project *Departments of Excellence* 2023-2027 awarded by the Ministry of University and Research). All authors contributed equally to conceiving the structure and content of this article. Andrea Pinotti wrote section 1; Federica Cavaletti wrote section 2; Pietro Conte wrote section 3 and conclusion.

analyses, the article proposes that VR's artistic and critical potential lies not in its capacity to simulate presence or evoke compassion, but in its ability to reveal its own mediological and performative construction. In this way, the essay contributes to reframing the debate on immersive media from a rhetoric of empathy and immediacy to one of awareness, opacity, and creative reflexivity.

1. The ambiguous dimension of virtual immersion as an empathy device

Although VR storytelling is still in search of a narrative form fully attuned to its technological affordances, its ›dispositive‹ – understood as the intertwining of a technical apparatus with the ideological discourses surrounding it – already positions itself as a catalyst for empathy (intended as bodily as well as affective resonance and capacity to put oneself in another's shoes), particularly in immersive projects that engage with global humanitarian crises. This potential becomes especially evident in documentary or docufiction works commonly grouped under the label of ›immersive journalism‹.[2]

Asked why she shifted from traditional cinema to VR for *The Protectors: Walk in the Ranger's Shoes* (a short 2017 documentary portraying a ranger's daily life in Garamba National Park, Democratic Republic of Congo),[3] Kathryn Bigelow answered without hesitation: »I think that the simple answer is empathy«.[4] Similarly, Ai Weiwei explicitly invokes empathy – tongqing 同情 in Chinese, »to share feelings or emotions with another person« – to frame the humanitarian scope of his work.[5] His VR piece *Omni* (actually a combination of two

2 See on empathy and VR Matthew Cotton (2021): Virtual Reality, Empathy and Ethics, Cham: Springer-Palgrave Macmillan. More specifically on VR and the humanitarian crisis see Francesco Zucconi (2018): Displacing Caravaggio. Art, Media, and Humanitarian Visual Culture, Cham: Palgrave Macmillan, Ch. 5 »On the limits of the virtual humanitarian experience«, pp. 149–181.

3 *The Protectors* (2017): »Walk in the Ranger's Shoes«, in: www.walkinginrangersshoes.com. Online: https://www.walkinginrangersshoes.com (last access: 13.10.2025).

4 Quoted in Adi Robertson (2017): »VR Was Sold as an ›Empathy Machine‹ – But Some Artists Are Getting Sick of It«, in: theverge.com (05.2017). Online: https://www.theverge.com/2017/5/3/15524404/tribeca-film-festival-2017-vr-empathy-machine-backlash (last access: 03.10.2025).

5 Ian Boyden (2018): »Not Yet Not Yet Complete. An Interview with Ai Weiwei. Part V: The Conditions of Empathy«, in: chinaheritage.net (10.2018). Online: http://chinaheri

separate videos: *Displaced Working Elephants in Myanmar* and *Rohingya Refugees in Bangladesh*) »gives viewers an intimate view of the uprooted, both animals and humans, as they experience various forms of displacement«,[6] allowing audiences to step into the place of both animals and refugees.

Chris Milk, when presenting his humanitarian VR productions for the United Nations, famously described VR as »the ultimate empathy machine«. In a well-known public talk, he emphasized his goal of drawing the spectator inside the »frame« of the story to provoke »more visceral emotional reactions«: »I don't want you in the frame, I want you *through* the frame, on the other side, in the world, inhabiting the world«.[7] For Milk, VR is an »experiential medium« – on the one hand a machine, on the other an experience that feels like »real life« or »truth«. The viewer feels physically present in that world and alongside the people represented within it.

Through the combined use of 3D cameras and binaural microphones, Milk achieves what might be described as the ›environmentalisation‹ of the image: a 360-degree visual and acoustic field in which the spectator experiences the sensation of ›being there‹. Watching *Clouds over Sidra*,[8] for example, you are not observing the Syrian girl protagonist through a ›screen‹ or ›window‹; you are sitting next to her in a refugee camp in Jordan, and »you empathize with her in a deeper way«.[9]

Implicitly, Milk suggests that VR environments promise to overcome a longstanding tradition – first articulated in the modern era by Leon Battista Alberti, who in his 1435 *On Painting* defined image perception as looking through a window at a fragment of the world and its actions (*historia*) unfolding within the frame: »I will say what I myself do when I paint. First, I trace as

tage.net/journal/the-conditions-of-empathy-ai-weiwei-interview-part-5/ (last access: 03.10.2025).

6 See Ai Weiwei, presented by The Guardian, in: acuteart.com. Online: https://www.acuteart.com/discover/ai-weiwei (last access: 03.10.2025).

7 Chris Milk (2015): »How Virtual Reality Can Create the Ultimate Empathy Machine« (TED Talk), online: https://www.youtube.com/watch?v=iXHil1TPxvA (last access: 03.10.2025).

8 The video is accessible also via browser at the following link: Within (27.01.2016, 0:08), Youtube: https://www.youtube.com/watch?v=mUosdCQsMkM&t=8s (last access: 03.10.2025).

9 Chris Milk (2015): »How Virtual Reality Can Create the Ultimate Empathy Machine« (TED Talk), online: https://www.youtube.com/watch?v=iXHil1TPxvA (last access: 03.10.2025).

large a quadrangle as I wish, with right angles, on the surface to be painted; in this place, it [the rectangular quadrangle] certainly functions for me as an open window through which the *historia* is observed«.[10] As has often been remarked, a thread connects Alberti's early modern reflection on painting to later technological developments, including the window-like display of our computer screens.[11] By contrast, a 360-degree immersive environment dispenses with frames altogether: the ability to look in all directions appears to liberate the viewer from the director's ›cut‹, collapsing framing onto the user's own gaze and allowing her to decide what to attend to and how to autonomously reframe her visual field.

Milk was highly optimistic about the anthropological, biopolitical, and humanitarian impact of this immersive mode. »We can change minds with this machine«,[12] he boldly proclaimed. His film was screened, for instance, at the World Economic Forum in Davos in January 2015 before policymakers whose decisions affect millions – and, according to him, these viewers »were affected« by Sidra's story. His collaboration with the United Nations[13] pursued the same aim: to present films directly to decision-makers capable of shaping the lives of those depicted. For Milk, this was only the beginning of discovering the »true power of virtual reality«[14] – namely, its capacity to alter how people perceive others and, ultimately, to transform the world.

A similar rationale has informed Alejandro González Iñárritu's work. In a BBC interview (2018), on the occasion of a Washington screening of his much-celebrated *Carne y Arena*, he elaborated on empathy and compassion as key emotions his VR installation seeks to elicit. Declaring himself uninterested in political activism and motivated solely by »humanity«, Iñárritu explained that his script was based on more than 120 migrant testimonies, aiming to enable spectators to share their experiences both sensorially and emotionally. But this sharing differs from the superficial connectedness associated with today's digital devices and social networks. Instead, Iñárritu calls for a solitary, first-person encounter between the VR user (isolated from her peripersonal

---

10 Leon Battista Alberti (2011): On Painting, Cambridge: Cambridge University Press, p. 55.

11 See Anne Friedberg (2006): The Virtual Window: from Alberti to Microsoft, Cambridge, MA: The MIT Press.

12 C. Milk: How Virtual Reality Can Create the Ultimate Empathy Machine.

13 See the presentation of this cooperation at https://www.youtube.com/watch?v=m-SZzyRFKK8 (last access: 03.10.2025).

14 C. Milk: How Virtual Reality Can Create the Ultimate Empathy Machine.

environment by the headset) and the migrants. To enforce this, participants must surrender their mobile phones at the entrance, preventing any real-time sharing during the 6.5-minute experience. As an immigrant himself, Iñárritu felt compelled to create this VR work in response to what he calls a »deficit of compassion« – a phrase that resonates with Barack Obama's 2006 speech on America's »empathy deficit« to graduates at Northwestern University.[15] This perspective adopts a restrictive notion of ›empathy‹ – equated with pro-social compassion, a humanitarian sympathy for the marginalized, the poor, the excluded. Yet empathy is not inherently benevolent and does possess its dark side: even sadists must be empathetic to better inflict pain. Watching the installation at its 2017 premiere at Fondazione Prada in Milan,[16] one might have wondered whether a right-wing Italian politician hostile to irregular migration would not instead empathize with the police patrol and their dogs...

Despite the shared ambition of Bigelow, Weiwei, Milk, and Iñárritu – to provoke powerful empathetic responses – their works differ in the Degrees of Freedom (DoF) offered by their VR technologies. The first three rely on 3DoF, while Iñárritu employs 6DoF. Borrowed from robotics, these terms distinguish environments where users may only orient their gaze across a 360-degree visual field (3DoF) from those where they may also move physically within the virtual space, approaching or retreating from digital entities (6DoF). Two further distinctions should be added: first, whether users are present solely as disembodied gazes or also represented through avatars – either »full-body« or »partial« (typically the hand); second, whether users remain passive spectators or are able to interact with the environment. In *Carne y Arena*, for example, the user can move within the 3D space but cannot interact with the virtual characters or objects.

Clearly, engagement intensifies when progressing from 3DoF to 6DoF and then to interactive 6DoF. Yet in all these immersive contexts, a paradox emerges: the ideological orientation of empathy is at odds with the supposed freedom enabled by 360-degree shooting. On the surface, the user's ability

15 »There's a lot of talk in this country about the federal deficit. But I think we should talk more about our empathy deficit [...]. We live in a culture that discourages empathy«: http://obamaspeeches.com/079-Northwestern-University-Commencement-Address-Obama-Speech.htm (last access: 03.10.2025).

16 Alejandro Iñárritu (2017): »Carne y Arena«, in: www.fondazioneprada.org. Online: https://www.fondazioneprada.org/project/carne-y-arena/?lang=en (last access: 03.10.2025).

to choose her perspective emancipates her from both the ›window‹ of Alberti and the director's framing. But at a deeper level, her gaze is subsumed into a predetermined emotional and ideological framework, nudged into reproducing the camera's sympathetic viewpoint. The more freedom the user feels in exploring an allegedly ›unframed‹ space,[17] the more insidiously the ideological conditioning asserts itself. Believing she acts autonomously, she in fact complies with a constraining framework composed by the authorial script, the affective preorientation, the predetermined possibilities of interaction and the constitutive limits of motor exploration imposed by the device.

This condition stems from the medium itself, which destabilises the conventional structure of image reception. In immersive VR, the observer's ability to distinguish support from image – the »twofoldness« that Richard Wollheim deemed essential to pictorial experience[18] – gives way to a blurring of image-consciousness and lived experience. In phenomenological terms, *Perzeption* (the perception proper to images, which does not posit its objects as real) yields to *Wahrnehmung* (literally, »taking as true«), the perception of things in their presence[19] The medium strives toward »transparent immediacy«:[20] the illusion of unmediated experience, achieved precisely through sophisticated mediation.

In this condition of transparency, spectators lose awareness not only of the technical construction of the scene but also of the director's perspective embedded within it. While believing themselves to be framing the world, they are in fact framed by the director's discourse. This loss of awareness dovetails with a further risk embedded in the rhetoric of virtual empathy:

---

17 *Contra*, see Mark Andrejevic/Zala Volcic (2019): »Virtual Empathy«, in: Communication, Culture & Critique 13/3, pp. 295–310. The authors call for a vindication both of the »linguistic« (i.e. mediated) character of the representation and its framedness: »Frames are incorrectly framed as an obstacle to perception, imagination, and empathy: rather, they are the conditions of possibility for all of these«. On framedness and framelessness see Pietro Conte (2020): Unframing Aesthetics, Milan/Udine: Mimesis International. On the rhetoric of unframedness see also: Anna Caterina Dalmasso (2019): »The Body as Virtual Frame. Performativity of the Image in Immersive Environments«, in: Cinéma & Cie 19/32, pp. 101–119.

18 See Richard Wollheim (2015): »Seeing-As, Seeing-In, and Pictorial Representation«, in: Id., Art and Its Objects, Cambridge: Cambridge University Press, pp. 137–151.

19 For this terminological distinction, see Edmund Husserl (2005 [1898–1925]): »Phantasy, Image Consciousness, and Memory«, Dordrecht: Springer, pp. 556 and 584.

20 Jay David Bolter/Richard Grusin (2000): Remediation. Understanding New Media, Cambridge, MA: MIT Press, p. 23.

the illusion of accessing the Other's experience directly, without recognizing its irreducible alterity.[21] Genuine empathy requires precisely such recognition, without fusing the Other into oneself. Instead, VR risks solipsistically absorbing the Other into the user's own perspective: »Technologies designed to foster empathy presume to acknowledge the experience of another, but inherently cannot. The user of these technologies, instead of acknowledging another's experience, hastily absorbs the other's experience into their own experience«.[22]

In this sense, ›perspective taking‹[23] – conceived as decentring oneself to inhabit another's viewpoint – devolves into projecting one's own subjectivity onto the virtual Other, who cannot reciprocate. I am »there«, but the Other's gaze remains blind, offering no recognition.

Again, this limitation is tied to the immersive apparatus itself. While 360° cameras promise unfettered spatial exploration and the illusion of comprehensively knowing the people depicted, such »totality« is illusory. Sidra's speech, the rangers' gestures, the Rohingya refugees' movements, the Mexican migrants' expressions – all remain fixed recordings, marked by what Roland Barthes famously described as photography's »that-has-been«.[24] The immersive setting may produce a powerful sense of ›being there‹, but it cannot yield new access to the emotional or existential depths of its subjects. Only an actual intersubjective relationship can provide such progressive disclosure. Not to mention the need to take into account the situated and implicated dimension of the actors involved: authors, users, actors, characters, in a perspective that,

---

21 On the necessity to separate a veritable empathy respecting the Other's Otherness from an undifferentiated fusion between subjects see the classic phenomenological perspectives developed by: Edith Stein (1964): On the Problem of Empathy, The Hague: Springer; Max Scheler (1954): The Nature of Sympathy, London: Routledge.

22 Grant Bollmer (2017): »Empathy Machines«, in: Media International Australia 165, pp. 63–78, here: p. 64. The same argument in Grant Bollmer (2020): »From Immersion to Empathy. The Legacy of Einfühlung in Virtual Reality and Digital Art«, in: Hava Aldouby (ed.), Shifting Identities. An Anthology of Presence, Empathy, and Agency in 21st Century Media Arts, Leuven: Leuven University Press, pp. 17–30, here: p. 26.

23 On the interpretation of empathy as ›perspective taking‹ see: C. Daniel Batson/Shannon Early/Giovanni Salvarani (1997): »Perspective Taking. Imagining How Another Feels Versus Imagining How You Would Feel«, in: Personality and Social Psychology Bulletin 23, pp. 751–758.

24 Roland Barthes (1981): Camera Lucida. Reflections on Photography, New York: Hill and Wang, pp. 76–77.

given the themes addressed by the works just mentioned, cannot ignore a decolonial perspective.[25]

Ultimately, otherness is doubly effaced: first, the director's gaze vanishes through the apparent transparency of the medium, leaving behind the inert corpse of reality; second, the represented subjects, though seemingly alive and mobile, are frozen like puppets, unable to respond. The first-person ›autopsy‹ of VR – *autos opsis*, »I have seen for myself« – thus shades into the ›autopsy‹ of a post-mortem, where life has already been stilled.

What has been described so far corresponds to an attempt to highlight the structural limitations of immersive VR technology aimed at eliciting empathetic feelings and embodied engagement in viewers: such shortcomings appear particularly evident when VR works parasite narrative and stylistic conventions imported by previous media, film and TV documentaries in the first. Let us now see in the next two paragraphs, dedicated to two specific case studies, whether a design and storytelling operation that critically and even subversively addresses the conventional structures of VR can effectively overcome these weaknesses, opening up new perspectives of mediological and experiential meaning.

2. Is anyone ›there‹? Unveiling the medium in *Wish You Were Here*

Among the aspects often associated with empathy and VR documentaries is the possibility to experience events ›directly‹. This can be interpreted according to two main meanings, corresponding to as many VR-specific ›illusions‹: experiencing events directly in the sense of ›with no mediation‹, which is associated with what is called »transparency illusion«; and in the sense of experiencing these events by ›being there‹, which is connected to what goes by the name of »place illusion«.

Following Jay D. Bolter and Diane Gromala, transparency illusion can be defined as the illusion that the media interface conveying images (and sounds)

25 See on this point: Patrick Colm Hogan (2015): »The Psychology of Colonialism and Postcolonialism: Cognitive Approaches to Identity and Empathy«, in: Liza Zunshine (ed.), The Oxford Handbook of Cognitive Literary Studies, Oxford: Oxford University Press, pp. 329–346.

can become invisible.[26] Such illusion is grounded in the designers' deliberate effort to hide the media interface, and sometimes in photorealism or even hyperrealism, i.e. the fact that the represented content is reproduced as realistically as possible, up to an ideally complete correspondence. The alleged consequence of transparency illusion is the perception that there is no mediation between us and what we see in the VR experience.

On its side, following this time Mel Slater and colleagues, place illusion can be described as the illusion of being in the place depicted in VR (›being there‹) in spite of the sure knowledge that this is not the case.[27] This illusion is grounded in a combination of factors. First, the fact that the VR content surrounds the spectator or user completely. Second, the fact that experiencing VR affords what Alva Noë and J. Kevin O'Regan call natural sensorimotor contingencies: that is, perception and action seemingly work like they do in a natural environment.[28] This also gives rise to a related and additional illusion, called »plausibility illusion«, which induces to take what happens in the virtual world as ›actually‹ happening. This phenomenon is more relevant in 6DoF than 3DoF environments, as it is more pronounced when a certain degree of interactivity is involved – which is the case for many, though not all, 6DoF environments. Lastly, place illusion is based on transparency illusion itself. Indeed, the idea is that, if there's no medium, then the content we experience is just here and now for us to plunge into it.

Importantly, all constructs just described are simply what their name implies: illusions. As for transparency illusion, the media interface is actually still there, even though possibly less evident. Far from being un-mediated, the depicted events are subject to at least two forms of mediation: the literal one of the interface; and the ideological one of the filmmaker's choices as to what to show and how.

Concerning place illusion, obviously enough, we are not actually transported anywhere. This is particularly worth stressing in the case of immersive

26 Jay David Bolter/Diane Gromala (2003): Windows and Mirrors. Interaction Design, Digital Art, and the Myth of Transparency, Cambridge, MA: The MIT Press.

27 Mel Slater (2009): »Place Illusion and Plausibility Can Lead to Realistic Behaviour in Immersive Virtual Environments«, in: Philosophical Transactions of the Royal Society B. Biological Sciences 364/1535, pp. 3549–3557; Mel Slater et al. (2022): »A Separate Reality: An Update on Place Illusion and Plausibility in Virtual Reality«, in: Frontiers in Virtual Reality 3/914392.

28 J. Kevin O'Regan/Alva Noë (2001): »A Sensorimotor Account of Vision and Visual Consciousness«, in: Behavioral and Brain Sciences 24/5, pp. 939–973.

documentaries, which often play with the idea of bringing the spectators to experience ›in first person‹ what is represented. Far from ›being there‹, together with the people being portrayed, we are removed from them both in space and time. In space, because – however illusionistic the representation can be – we are still in our real surroundings, while the people portrayed are part of the virtual images; and in time, because normally the events we are witnessing are not broadcast ›live‹, but they have been recorded or designed early on.

By now, as far as the scholarly debate is concerned, it is generally well established that transparency illusion and place illusion are to be taken for what they are, to the point that their critique has probably become trite. However, overemphasising VR's ability to show things ›with no mediation‹, to ›bring people there‹ has remained a widespread and quite unquestioned strategy for marketing immersive documentaries.

In fact, this is functional for most parts involved. Filmmakers capitalize on the promise of an unusually powerful and involving experience; expert users, after being disappointed in the past, can keep their hopes high that ›this time‹ they will actually get such experience; on their side, naive users (and there is still plenty of them) may simply believe that they will get such experience. As a result, authors tend to ›fool‹ the audience, and spectators tend to be happy to be ›fooled‹.

This state of affairs risks having detrimental consequences for VR. First, it can create distrust and disappointment in the medium: after all, in the long run, not even an audience that is open to being tricked really likes not getting what they keep being promised. Second, it can transform VR contents into a mere pretext for VR *effects*. This is almost inevitable, if what counts the most is achieving the forms of illusion discussed above. Lastly, and for a very similar reason, it can generate repetitive products, because they basically all have the same underlying goal.

This set of consideration takes on particular importance when the object of the VR representation is, in many possible ways, *sensitive*, or when it shows conditions that are unfamiliar to the audience, with the intent of making them ›step into‹ the condition at stake in first person. A relevant field in this regard is the representation of *illness*.

VR directors have developed different strategies to avoid overemphasising unrealistically the illusionistic properties of the medium, and to stay away from the problematic and likely offensive assumption that simply going through a VR experience is enough to experience the depicted illness directly.

In some cases, the choice is made to create some distance from the representation by privileging CGI over filmic, photorealistic images. While photorealism is not a necessary condition for place illusion, it definitely plays a crucial role for transparency illusion. Plunging into a CGI world can give the impression of being in that world, but can hardly convey the sense that we are witnessing an event as it unfolds and with no mediation: life is not a cartoon. This has been the strategy adopted by notable representations of both organic and mental illnesses, based on real stories, like *Goliath. Playing with Reality* (Barry Gene Murphy & May Abdalla, 2019) narrating schizophrenia; and *Notes on Blindness* (Amaury La Burthe, Arnaud Colinart, James Spinney, Peter Middleton, 2016), on the experience of losing gradually one's sight.

However, this is not to conclude that using photorealistic images necessarily leads to illusionistic VR in the senses discussed above. To demonstrate how things can be done differently, a case study will now be introduced and analyzed more in depth.

Wish You Were Here (2023) is a 3DoF VR documentary by theatre, film, and VR director Michał Stankiewicz about Aneta, a 36-year-old woman with breast cancer. Spectators can follow different aspects of Aneta's life in her home, as well as listen to some of her reflections. They cannot interact, but they can explore what they see all around themselves.

When the documentary begins, what is shown is an orchard that grows opposite Aneta's house.[29] The woman lets spectators know that they will be watching a documentary in VR, and warns, somehow enigmatically, that the representation contains a lot of errors. Deliberately, nothing in this quite long starting scene suggests that we are ›there‹. In fact, it is already possible to foresee what is particular about *Wish You Were Here*. A microphone can be spotted in the middle of the vegetation in the first frame, and a cameraman with a camera becomes visible too in the second frame.

The next scene is set inside Aneta's flat. The spectators can see how she lives, and meet her two children, whom she raises alone. However, as soon as an illusion of presence may start to set in, the film crew shooting the documentary appears in the scene.

Later on, the documentary reveals that Aneta has cancer and lets the spectator listen to her reflections about it: how she perceives her condition as absurd, sometimes not even real; how she realises that the disease is eating away

29 Description was adapted from materials courtesy of the filmmaker.

at her quality of life day by day, and how she still tries to enjoy her remaining time as fully as she can.

From this moment on, spectators will be projected into different seasons by means of a series of flashforwards. In winter, they see Aneta in more intimate situations: as she plays with her daughter, but also as she gets a massage from a therapist to ease her pain and discomfort after mastectomy. Again, however, illusionistic immersion is hindered by the film crew being visible on the scene.

In spring, Aneta's conditions have worsened significantly, and she no longer has hair. Spectators are in her room, and nothing happens, except for Aneta at some point looking at the camera. The next scene is set in Aneta's bathroom, where she says she sometimes hides when she wants to cry without being seen or heard. Again, however, this very intimate confession, which may bring spectators illusionistically close to the woman, is counterbalanced by the camera being shown, very visibly, in the bathroom mirror and other reflecting surfaces.

At this point, the flow of the filmic images is interrupted. Aneta shares that when she needs to calm down she breaths deeply and imagines being immersed in blue. As she says so, the field of vision actually becomes blue.

Finally, spectators are projected in summer. They see the orchard and a large new building. Later, they are shown an empty room, with only a camera where Aneta was previously sitting.

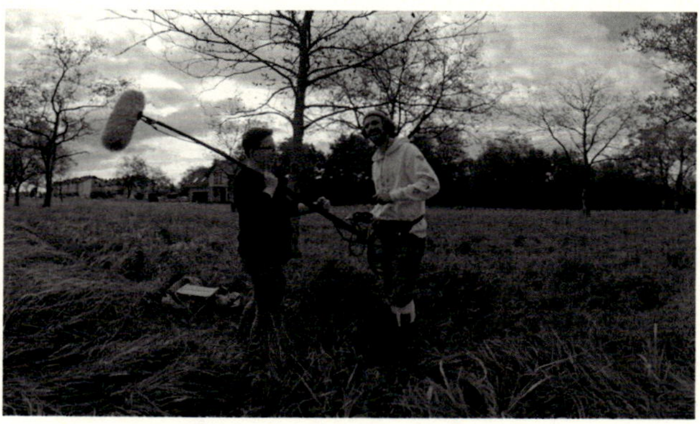

fig. 1: Wish You Were Here, 2023, VR documentary, R: Michał Stankiewicz. Aneta interacts playfully with a film crew member

As it is easy to conclude, the distinctive feature of *Wish You Were Here* is that, as opposed to common practice, the director does not hide the traces of the VR filmic production. On the contrary, he leaves the recording equipment in full sight, and sometimes he even artificially doubles or triples it. In the bathroom shot, for instance, the camera is reflected not only in one, but in several reflecting surfaces.

In addition, the protagonist herself acknowledges both the production equipment and the presence of the spectator(s). At the beginning of the documentary, she declares outright that »this is a film, a documentary«; that she is »walking around the camera right now«; and, most strikingly, that »you [spectators] keep following me with your eyes«. Even more than the cameras and staff in sight, this direct address is particularly impactful as it comes really early and suddenly in the documentary, catching spectators off-guard before they can understand that in the director's setting they will be ›seen‹.

What are the consequences of these choices? First of all, the unveiling of the production equipment and Aneta's meta-filmic attitude tell spectators unambiguously that what they see was recorded and, thus, the events are mediated: this is the end of transparency illusion. The measures adopted in this VR documentary against transparency, not so rare to be found in cinema too, hit somehow more strongly than if they were deployed in a non-immersive movie. This is because in that case, the margins of the screen (the film frame, so to speak) constitute an ongoing reminder of the mediated nature of the representation. As a consequence of the end of transparency illusion, place illusion cracks too: if events are mediated, then they cannot be ›accessed‹. In this case, this is primarily due to time-related considerations depending on the filmic nature of the images, and our awareness of the way they work: the mediation that we are called to acknowledge hints quite unambiguously at the fact that the events are not unfolding *now*, as we watch them, but they took place and were filmed in the past. We cannot access them any more than we can access a non-immersive movie. The director, therefore, refuses to exploit the two most ›classic‹ illusions of mainstream VR documentaries.

The director's awareness and the fact that the fall of transparency and place illusion are fully deliberate is confirmed by his own discussion of *Wish You Were Here*. In a 2022 interview, he clarifies how he decided to turn a VR-specific feature against itself. The reference is to the »unframedness« already mentioned in this article, i.e. the (once again) illusion that, as it envelops spectators in a 360° field, different from traditional types of images, VR representations extend the limits of their frame until they ideally suppress them. In mainstream

VR practice, unframedness is used to promote transparency illusion and place illusion: in fact, essentially, the absence of a frame implies the absence of a medium as well. Stankiewicz, however, sagaciously observes that, if the frame gets so wide that you can actually look all around, at some point one becomes able to see what is behind the scene: namely, the film production equipment and crew. The take-home-message to VR creators can thus be read as a sort of cautionary tale: VR strength can just as easily become its weakness.

However, so to speak, the director appears to have a message for VR spectators as well. To grasp this, it is useful to go at the roots of Stankiewicz's inspiration, which is undoubtedly Brechtian. Indeed, the visibility of the recording equipment, and even more the protagonist's direct gaze to the spectators are clearly to be traced back to Brecht's notion of breaking the fourth wall. This move, it is worth remembering, was aimed at creating more aware and critical spectators. In an oft-cited quote by the German playwright: »The spectator was no longer in any way allowed to submit to an experience uncritically (and without practical consequences) [...] The production took the subject-matter and the incidents shown and put them through a process of alienation: the alienation that is necessary to all understanding«.[30]

Making the VR experience less obvious, less seemingly natural, which is a responsibility of VR creators, implies in a specular way that spectators too engage in a deeper consideration of their role and even duties. If VR creators should stop relying easily on VR-specific illusions, then spectators should stop falling for them just as easily. This is particularly relevant for VR documentaries, so often showing »the pain of others«.[31] In *Wish You Were Here*, spectators are called even aggressively (»you keep following me with your eyes«) to take into account that they are being given (partial, mediated) access to something very sensitive. In the same interview, indeed, the director states that VR gives spectators the opportunity to see themselves in a way, and – most importantly – to ask *why* they are looking. Which is important to avoid not only mindless immersion, but also any form of voyeurism.

Before concluding this analysis of *Wish You Were Here*, it is worth adding some words about the moment in which, triggered by Aneta's confidence that

30 Bertolt Brecht (1974 [about 1936]): »Theatre for Pleasure or Theatre for Instruction«, in: John Willet (ed.), Brecht on Theatre. The Development of an Aesthetics, London: Eyre Methuen, pp. 69–76.

31 Susan Sontag (2003): Regarding the Pain of Others, New York: Farrar, Straus and Giroux.

when she wants to relax she imagines being in the blue, the visual field is concretely filled with blue. In the director's words, this time from a summary[32] of the documentary, until that moment the experience has been »cinematic« and »documentary«, but now it becomes »magical«. »It's a long immersive scene« the director adds explicitly, in which we get to »being together in Aneta's imaginary blue space«. Significantly, then, the most immersive moment of the documentary, when we get closest to the protagonist, almost to the point of being with her, is not illusionistic at all. This can be taken as yet another demonstration that alternatives to mainstream forms of alleged connection and intimacy are just there, available to be experimented with.

3. Live on stage: hacking authenticity in *Great Hoax: The Moon Landing*

While *Wish You Were Here* explicitly questions the assumed transparency and objectivity traditionally associated with the documentary form, it can still be classified as a documentary, as it ultimately does not dispute the authenticity of the narrative content it presents. In contrast, other VR experiences go beyond problematizing the myth of transparent narration to also question the truthfulness of the content itself, thereby prompting a deeper reflection on the role of belief – and, consequently, on the relationship between veridicity and fiction – in immersive storytelling.

Great Hoax: The Moon Landing is a 17-minute interactive VR comedy created in 2019 by John Hsu and Marco Lococo, which premiered at the 2020 Venice Film Festival's VR Expanded section. Designed as a room-scale experience providing six degrees of freedom, it enables users to move around within a designated play area, as tracking sensors capture their movements in all directions and translate them into the virtual environment in real time.

The plot is simple. Half a century after Apollo 11's 1969 moon landing showcased to the world the heights a powerful nation could reach, the Taiwanese people are grappling with deep-rooted identity struggles and growing financial instability. In search of renewed purpose, they yearn for a great national enterprise that could elevate the island's prestige on the global stage. The ideal solution? A moon landing. Yet with no real budget for space exploration, there is only one option left: to fake it.

32 Part of the directors' private portfolio. Access courtesy of the director.

This is where you step in. No matter where you actually come from, you are expected to empathise with the Taiwanese people (including VR creator John Hsu) and make every effort to help improve the island's socio-political and economic conditions. Yet soon, this form of empathy, initially appearing to align with the rhetoric common in VR documentaries, evolves into something entirely different – something with Hsu himself has defined as »political satire«.[33]

In the first scene, a barely credible film director asks if you are willing to »choose your destiny« and possibly become the first Taiwanese astronaut on the moon (fig. 2).

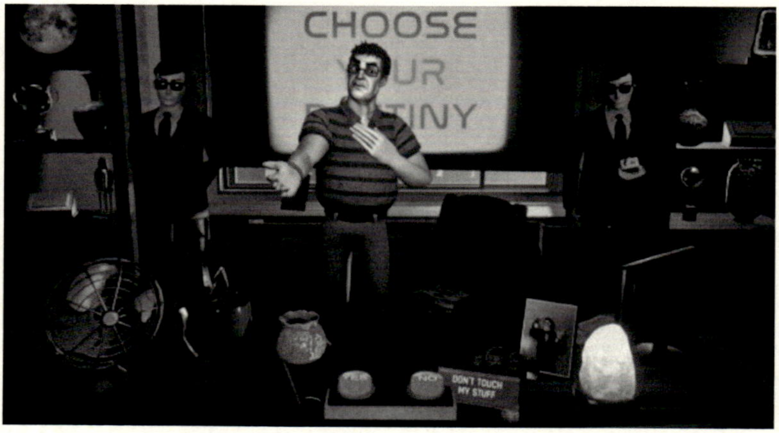

fig. 2: John Hsu, Marco Lococo: Great Hoax: The Moon Landing, 2019, Serendipity/3dar. The film director enlisting you to play the first Taiwanese astronaut on the moon

By pressing the »Yes« button on the left (thus disregarding the »Don't touch my stuff« sign on the desk right next to it), you are suddenly transported to a studio designed to resemble the Moon's surface, where a film crew follows

33 John Hsu/Marco Lococo (2020): »Directors' Statement«, in: www.labiennale.org. Online: https://www.labiennale.org/en/cinema/2020/venice-vr-expanded/great-hoax-moon-landing (last access: 9.10.2025).

the director's orders and a pair of men, who look suspiciously like government agents, oversee the entire operation. Most strikingly, as you step out of the moon capsule, you catch your own reflection in a mirror-like screen directly in front of you – a classic strategy designed to make you feel embodied in the avatar and present in the virtual world (fig. 3).[34]

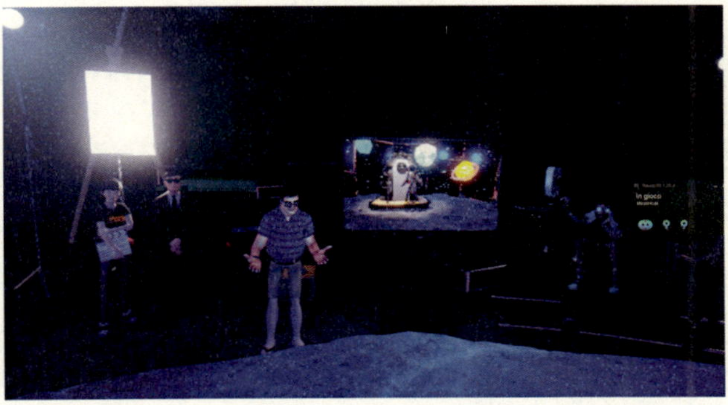

fig. 3: John Hsu, Marco Lococo: Great Hoax: The Moon Landing, 2019, Serendipity/3dar. On set, filming the staged moon landing

Following the director's instructions, you strive to deliver your most convincing performance. Since the entire experience is rendered in real-time CGI, you can slowly step out of the space capsule and take that famous »small step for man, giant leap for mankind«. You can then plant a flag on the lunar surface and carry out all the heroic or pseudo-heroic actions and poses you are

34 Mar Gonzalez-Franco/Daniel Perez-Marcos/Bernhard Spanlang/Mel Slater (2010): »The contribution of real-time mirror reflections of motor actions on virtual body ownership in an immersive virtual environment«, in: 2010 IEEE Virtual Reality Conference (VR), Waltham, MA: Curran Associates, pp. 111–114, https://doi.org/10.1109/VR.2010.5444805 ; Yasuyuki Inoue/Michiteru Kitazaki (2021): »Virtual Mirror and Beyond: The Psychological Basis for Avatar Embodiment via a Mirror«, in: Journal of Robotics and Mechatronics 33/5, pp. 1004–1012; Çağlar Yildirim/Sercan Şengün/Mehmet Akhoroz/Eyup Kucuk/Fox Harrell (2023): »Through the Looking Glass. The Role of Virtual Mirrors in Shaping Empathy in Virtual Reality Perspective Taking«, in: Proceedings of the 22nd International Conference on Mobile and Ubiquitous Multimedia, pp. 501–504, https://doi.org/10.1145/3626705.3631789.

asked to perform, or simply those you choose, guided solely by your own will. Once the shoot is finished, the next scene takes you back to your small, dimly lit apartment, where you are counting down the days until the release date. The newspaper announces that the moon landing was successful. Finally, the phone rings, and it is once again the director, asking you to watch the TV broadcast of the epic undertaking. At first, everything seems fine, with two anchors breaking the news of the momentous event. Then you realise that the end product is no longer about the moon landing but has taken an unexpected and ironic turn: the ›original‹ footage that you helped to create has been repurposed as a kitsch Taiwanese karaoke music video with the light-hearted style of a propaganda piece. The gestures and movements you performed earlier now appear as choreographed, yet clumsy dance steps set to a patriotic pop tune (fig. 4).

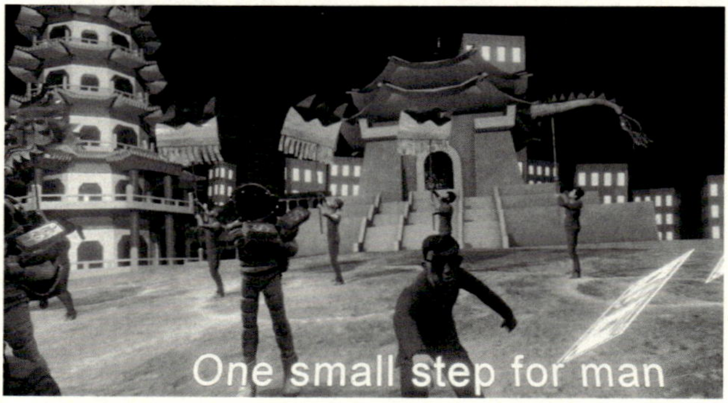

fig. 4: John Hsu, Marco Lococo: Great Hoax: The Moon Landing, 2019, Serendipity/3dar. Your ›original‹ performance is edited into a music video

As fireworks illuminate the night sky above the island, an overwhelming wave of anger compels you to explore the interactive possibilities the simulation affords, like tossing objects out the window or hurling an ashtray at the television, shattering its screen. You feel completely alone until, in the final scene, your surroundings begin to dissolve, leaving you hallucinating, stranded once again on the (fake?) moon. This time, you are all by yourself, staring back at the Earth in silence. Eventually, adding insult to injury, you are presented with a fine for ridiculous charges, including an »incompetent acting« fee and a »Director's psychiatric bill«.

As in a classical movie, the experience ends with the credits. Yet this was not a film but an immersive VR experience. The first question, then, is: how should we define this work? From the outset, you – the user – are asked to put yourself in someone else's shoes. This »someone else«, however, is an *actor*. Contrary to typical three-degrees-of-freedom VR documentaries such as those found in so-called immersive journalism, which claim to teleport you to another place and make you believe you are experiencing someone else's real life, *Great Hoax* does not position you as a mere spectator confronted with a recording and its supposed truthfulness. Instead, you are invited to play a game, perform a role, and act like a true actor, thereby contributing to the creation of a fake – that is, to convey the impression of a documentary that is not really a documentary but a fabrication. In this particular case, suspension of disbelief is triggered right away, which neutralises any attempt to generate plausibility or credibility. If we were to interpret *Great Hoax* from a phenomenological point of view, we might be tempted to say that an original consciousness involving position-taking (that is, believing or disbelieving in what is being shown) comes to be modified and replaced by a neutralised act in which positionality is suppressed, thereby bringing suspension of (dis)belief increasingly into focus.[35]

This point, however, warrants further clarification. The lack of plausibility pertains to the *diegetic* world – that is, to the story in which you are asked to become the main character, the hero. Yet there is a second level of credibility: one that pertains to *you* as the individual in the flesh wearing the headset. You are asked to perform convincingly, making the disembarking from the spaceship and your every action seem believable to any potential future viewers. While you remain fully aware of the fictional, non-positional nature of the ›movie‹ *within* the VR experience, you nonetheless aim to generate a kind of credibility *within* fictionality – a paradoxical credibility *of* fictionality.

References to the most famous ›real‹ (if one might define it so) ›great hoax‹ come into play here. Conspiracy theorists use this term to refer to the 1969 moon landing, which they claim was faked. A direct reference to this narrative can be seen in the title itself of the VR experience. However, adopting a media-archaeological perspective reveals that the idea of a moon-related hoax predates Apollo 11 by more than a century. In 1835, the *New York Sun* published

35 See Edmund Husserl (2005 [1922-1923]): »Pure Possibility and Phantasy«, in: Id., Phantasy, Image Consciousness, and Memory, Dordrecht: Springer, pp. 659–677.

a series of six articles claiming the discovery of life on the moon. Collectively known as *The Great Moon Hoax*, these articles represent an early instance of science fiction being presented as factual news. The articles were allegedly reprinted from the *Edinburgh Journal of Science*, with authorship attributed to a Dr Andrew Grant, described as a colleague of John Herschel, one of the most respected astronomers of the time. Herschel had indeed travelled to Cape Town, South Africa, in January 1834 to set up an observatory equipped with a powerful new telescope. According to Grant's fictional account, Herschel had observed life forms on the moon, including fantastical creatures such as unicorns and winged humanoids resembling bats.

To be sure, the story was a complete fabrication. The *Edinburgh Journal of Science* had ceased publication several years earlier, and Dr Andrew Grant was a fictional character. Intended as satire, the articles were crafted to mock earlier, earnest speculations about extraterrestrial life. However, readers at the time were most often deceived: they failed to recognize the story as a parody. The excitement over Herschel's alleged discoveries was so widespread that even a committee of scientists from Yale University travelled to New York in search of the original *Edinburgh Journal* articles. It was only weeks later that the *Sun* confessed the entire story had been a hoax.

In the VR experience, a reference to this story appears in the newspaper on the table of our cluttered room. There is also a second direct reference to a more recent and famous chapter of the great hoax, that is, the hypothesis that the Apollo 11 moon landing itself was fake news and that the entire documentary was staged. Several details clearly hint at this, particularly the version of the conspiracy theory claiming that Stanley Kubrick himself was the mastermind behind the fabricated landing. The theory began with a straightforward question: who in 1969 could have realistically staged a convincing moon landing? Kubrick's masterpiece, *2001: A Space Odyssey*, had been released just the year before. From this starting point, the theory evolved, becoming increasingly elaborate and incorporating progressively more bizarre claims.[36]

36 The theory was initially proposed in Bill Kaysing (1976): We never went to the moon, Pomeroy: Health Research. For more on this, see Roger Launius (2010): »Denying the Apollo Moon Landings: Conspiracy and Questioning in Modern American History«, in: 48th AIAA Aerospace Sciences Meeting Including the New Horizons Forum and Aerospace Exposition, https://doi.org/10.2514/6.2010-1131. For a critique of the main moon hoax conspiracy theories, see Paolo Attivissimo (2013): Moon Hoax, Debunked!, Raleigh: Lulu.com.

The VR work references all of this primarily through its soundtrack, which features the most iconic musical theme from *2001*, Richard Strauss's *Also sprach Zarathustra*, and through the name of the film director Stanley King, a blend of Stanley Kubrick and Stephen King which alludes to *The Shining*, often interpreted by conspiracy theorists as Kubrick's admission of involvement in the fake moon landing.[37]

Let us now return to our central question: how should we define the VR experience? This is not merely a case of VR being used to present a fabricated video or a fictional documentary. The situation becomes more complex when one begins to notice specific details that challenge, and ultimately subvert, the fictional status of the (alleged) ›fake‹ documentary. This is particularly evident in paradigmatic moments, such as when the viewer looks down at their space suit and notices that it is covered in patches – an implausible feature in the context of an authentic astronaut equipment. Or, similarly, when one observes their hands only to find that they are wearing comical kitchen gloves adorned with cartoonish kittens (fig. 5).

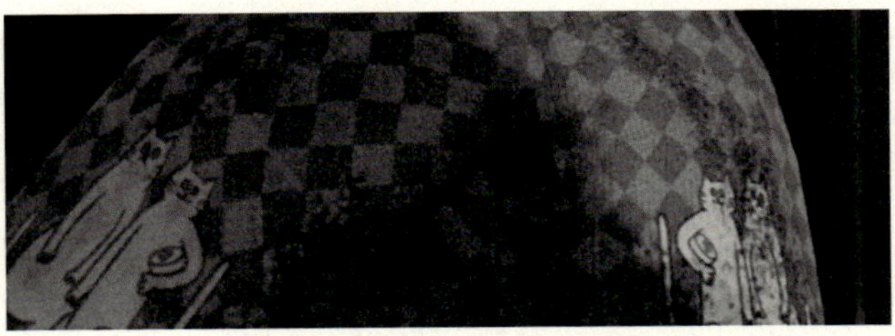

fig. 5: John Hsu, Marco Lococo: Great Hoax: The Moon Landing, 2019, Serendipity/3dar. Hardly standard astronaut equipment

37 Lunar conspiracy theories that place Kubrick at the centre of the plot have been the subject of numerous films, including William Karel's *Opération Lune/Dark Side of the Moon* (2002), Rodney Ascher's *Room 237* (2012), Antoine Bardou-Jacquet's *Moonwalkers* (2015), and Fabrice Mathieu's *Moon Shining*, which was created by combining footage from the Apollo 11 mission with several authentic Stanley Kubrick interviews. Similar conspiracy themes have also been explored in relation to Mars, most notably in Peter Hyams' *Capricorn One* (1977).

At such junctures, the user's apprehension, initially aligned with documentary codes and conventions, shifts toward mockumentary aesthetics, marked by irony, self-reflexivity, and a challenging of factual evidence and veridictive marks. Those deliberately inserted elements are designed to produce a sense of scepticism in the user, drawing attention to the constructed nature of the narrative and signalling the actual role they are *expected* to assume. It becomes increasingly clear that the user is not, in fact, meant to experience the story as the first Taiwanese astronaut on the moon, but rather to recognise and engage with a more complex interpretative role shaped by an awareness of narrative construction, epistemological questioning, and a heightened sense of mediation.

However, this is not enough since, at the end of the experience, you see your own performance transformed into a music video, which is fictional – *explicitly* fictional – by definition. This occurs without your consent, as often happens in contemporary deepfake videos. The motif of constructed reality and fabricated authenticity resurfaces, but unlike at the beginning of the experience, it is no longer you deceiving the viewers; rather, you are the one being manipulated – faked – by the director and his crew. Seen from this perspective, *Great Hoax* proves to be not only a remediation of previous media (newspaper, cinema, TV), but also a meta-reflection on VR in general and its relationship to documentary cinema in particular.

Indeed, while the whole cinematic apparatus is overtly displayed within the immersive digital environment, thus revealing the opacity of the film medium, things are quite different in relation to the medium of VR. First-person perspective, real-time body tracking, six degrees of freedom, and the interactive structure of the narrative combine to generate a strong feeling of presence, thereby inducing place illusion, plausibility illusion, and the illusion of ownership over the virtual body. This, in turn, contributes to conveying the impression of transparency of the technology. And yet, as I argue, the ironic conclusion of the diegetic experience, where your efforts to lend credibility to the fake moon landing are transformed into a parody, is at the same time a self-reflexive gesture that prompts you, the experiencer in the flesh, to reflect on the opacity of the VR medium. You are invited to leave the story, so to speak; that is, to exit the diegetic world and meditate on the artefactual, hypermediated nature of the extra-diegetic world – namely, the VR experience itself.

Moreover, as soon as you abstract yourself from the engaging narrative of the experience, you may begin to reflect on the opacity not only of the medium but also of its underlying technological apparatus. Tracking is among the most crucial technical characteristics of VR, playing a key role in defining its medium specificity. In 6DoF experiences, in particular, real-time tracking of the user's body movements – and, in some cases, facial expressions – is constantly processed by algorithms that dynamically adjust the virtual environment in response, enhancing immersion and (potentially) interactivity. Beyond this ›attunement‹ between the environment and the user, however, tracked movements can be (and often are) stored, temporarily or persistently, for multifarious purposes: from simply saving the user's position in a game, to personalising experiences, training AI models (for example, to enhance avatar animation or gaze prediction), or even enabling behavioural profiling. This means that data resulting from tracking can be used without the user knowledge – let alone their consent. Here is where the imaginary story of *Great Hoax* and factual reality converge in a shared opacity. At the beginning of the experience, when you believed you had to fake a Moon landing, *you* were in control. At the end, however, you realise you have been deceived, and you also come to understand that your image has been exploited to create something that completely exceeds your control – making you aware that you yourself are being controlled. In this sense, *Great Hoax* is also a VR experience that urges users to reflect on the potential (mis)uses of their personal image and movements, and on the risk that their identity could be repurposed in ways that serve interests other than their own.

Disguised as a funny and rather naïve experiment, *Great Hoax* reveals itself to be a serious and thought-provoking VR experience. It urges the user to reflect on the nature of VR as a medium, to engage critically with one of its most popular, acclaimed, and ideologically charged genres – immersive journalism – and its associated truth claims, and to adopt a perspective aimed at debunking the rhetoric and hyperbolic assertions that often accompany it. All things considered, it is precisely the meta-critical awareness that *Great Hoax* elicits that makes it an artistic, rather than merely ludic, experience.

Conclusion

Through the lens of *Wish You Were Here* and *Great Hoax: The Moon Landing*, this essay has aimed not only to contribute to the debate surrounding the bombastic rhetoric of VR as an empathy machine, but also to reveal that the myth of VR's transparency is far more nuanced than commonly assumed. Crucially, this very complexity can be harnessed to create medium-specific artistic experiences, as demonstrated in the aforementioned works, which engage in meta-reflection on the ambiguous strategies used to evoke the feeling of »being there«, as well as on the role and potential misuse of tracking technologies which are central to both 3DoF and 6DoF VR environments.

In *Wish You Were Here*, the illusion of seamless immersion is deliberately broken by exposing the documentary crew at work, forcing viewers to confront the constructed (and in this sense opaque) nature of the VR medium. This strategy directly challenges the purported immediacy and objectivity often claimed by immersive journalism.

Great Hoax pushes this critique even further by destabilising not only the transparency of the medium but also the truthfulness of the narrative content it vehiculates. Being invited to participate in a fabricated documentary, we are initially positioned to perform credibility within the story. As the experience unfolds, however, we become increasingly aware of the performative nature not only of the infra-diegetic world of the fake documentary, but also of what we are asked to do *as VR users*. This paradoxically draws attention not only to the constructed nature of the documentary we are helping to create, but also to the artificiality of the VR experience we are undergoing as physical participants. Such multilayered construction unsettles our positionality, exposing the inherent tensions between belief, fiction, manipulation, and the VR medium itself. The experience highlights how VR's power can indeed lie not (only) in transparent and empathy-inducing truth, but (also) in the capacity to make users critically aware of their role in co-producing engaging narratives.

Indeed, both case studies examined prompt a reflection on how VR continues to be described as the ultimate medium for eliciting empathy and generating a sense of presence in users. *Wish You Were Here* appears to reintroduce the traditional notion of empathy as the act of feeling someone else's emotions (specifically, their pain). Yet, the overt display of the technological apparatus abruptly breaks the illusion of presence and, with it, an overly simplistic understanding of empathy. However, this disruption does not discredit empathy itself. On the contrary, the artistic choice encourages users to reflect on the fact that their presence is only a simulated one – an important consideration when realizing that they are not truly ›there‹ with the terminally ill patient. Empathy, in this context, is not evoked through an uncritical belief in the medium's supposed ›magical‹ power, but rather through an awareness of its inherent limitations. It is precisely this acknowledgment that deepens and enriches the empathetic experience.

In *Great Hoax*, the naïve acceptance of the rhetoric surrounding VR is challenged through the adoption of a ludic stance, in which users are invited to step into the role of an actor or player within a game-like environment. Rather than empathizing with the Taiwanese people in their struggle for national grandeur, participants are instead prompted to reflect on the nature of propaganda and mystification – including those tied to the very technological apparatus they are using (and by which they are, in turn, being used).

Ultimately, these examples underscore the necessity of approaching VR storytelling with a critical eye. Rather than accepting the myth of VR as an objective, transparent ›looking glass‹ into others' experiences, we must acknowledge the complex interplay between narrative construction, user performativity, and the awareness of the artefactual nature of the medium. It is precisely this kind of meta-reflection – one that playfully turns VR against itself – that can help bringing the *artistic* possibilities of VR to the fore.

Bibliography

Alberti, Leon Battista (2011): On Painting, Cambridge: Cambridge University Press.

Andrejevic, Mark/Zala Volcic (2019): »Virtual Empathy«, in: Communication, Culture & Critique 13/3, pp. 295–310.

Attivissimo, Paolo (2013): Moon Hoax, Debunked!, Raleigh: Lulu.com.

Barthes, Roland (1981): Camera Lucida. Reflections on Photography, New York: Hill and Wang.

Batson, C. Daniel/Shannon Early/Giovanni Salvarani (1997): »Perspective Taking. Imagining How Another Feels Versus Imagining How You Would Feel«, in: Personality and Social Psychology Bulletin 23, pp. 751–758.

Bollmer, Grant (2017): »Empathy Machines«, in: Media International Australia 165, pp. 63–78.

Bollmer, Grant (2020): »From Immersion to Empathy. The Legacy of Einfühlung in Virtual Reality and Digital Art«, in: Hava Aldouby (ed.), Shifting Identities. An Anthology of Presence, Empathy, and Agency in 21st Century Media Arts, Leuven: Leuven University Press, pp. 17–30.

Bolter, Jay David/Diane Gromala (2003): Windows and Mirrors: Interaction Design, Digital Art, and the Myth of Transparency, Cambridge, MA: The MIT Press.

Bolter, Jay David/Richard Grusin (2000): Remediation: Understanding New Media, Cambridge, MA: MIT Press.

Boyden, Ian (2018): »Not Yet Not Yet Complete. An Interview with Ai Weiwei. Part V: The Conditions of Empathy«, in: chinaheritage.net (10.2018). Online: http://chinaheritage.net/journal/the-conditions-of-empathy-ai-weiwei-interview-part-5/ (last access: 03.10.2025).

Brecht, Bertolt (1974 [about 1936]): »Theatre for Pleasure or Theatre for Instruction«, in: John Willet (ed.), Brecht on Theatre. The Development of an Aesthetics, London: Eyre Methuen, pp. 69–76.

Conte, Pietro (2020): Unframing Aesthetics, Milan-Udine: Mimesis International 2020.

Cotton, Matthew (2021): Virtual Reality, Empathy and Ethics, Cham: Springer-Palgrave Macmillan.

Dalmasso, Anna Caterina (2019): »The Body as Virtual Frame. Performativity of the Image in Immersive Environments«, in: Cinéma & Cie 19/32, pp. 101–119.

Formenti, Cristina (2014): Il mockumentary. La fiction si maschera da documentario, Milan/Udine: Mimesis.

Friedberg, Anne (2006): The Virtual Window: from Alberti to Microsoft. Cambridge, MA: The MIT Press.

Gonzalez-Franco, Mar/Daniel Perez-Marcos/Bernhard Spanlang/Mel Slater (2010): »The contribution of real-time mirror reflections of motor actions on virtual body ownership in an immersive virtual environment«, in: 2010 IEEE Virtual Reality Conference (VR), Waltham, MA: Curran Associates, pp. 111–114. DOI:10.1109/VR.2010.5444805 (last access: 13.10.2015)

Hsu, John/Marco Lococo (2020): »Directors' Statement«, in: www.labiennale.org. Online: https://www.labiennale.org/en/cinema/2020/venice-vr-expanded/great-hoax-moon-landing (last access: 9.10.2025).

Husserl, Edmund (2005 [1898-1925]): Phantasy, Image Consciousness, and Memory, Springer: Dordrecht.

Husserl, Edmund (2005 [1922-1923]): »Pure Possibility and Phantasy«, in: Id., Phantasy, Image Consciousness, and Memory, Springer: Dordrecht, pp. 659–677.

Inoue, Yasuyuki/Michiteru Kitazaki (2021): »Virtual Mirror and Beyond: The Psychological Basis for Avatar Embodiment via a Mirror«, in: Journal of Robotics and Mechatronics 33/5, pp. 1004–1012.

Kaysing, Bill (1976): We never went to the moon, Pomeroy: Health Research.

Launius, Roger (2010): »Denying the Apollo Moon Landings: Conspiracy and Questioning in Modern American History«, in: 48th AIAA Aerospace Sciences Meeting Including the New Horizons Forum and Aerospace Exposition. DOI: https://doi.org/10.2514/6.2010-1131

Miller, Cynthia J. (ed.) (2012): Too Bold for the Box Office: The Mockumentary from Big Screen to Small, Lanham, MD: Scarecrow Press.

Milk, Chris (2015): »How Virtual Reality Can Create the Ultimate Empathy Machine« (TED Talk), online: https://www.youtube.com/watch?v=iXHil1TPxvA (last access: 03.10.2025).

O'Regan, J. Kevin/Alva Noë (2001): »A Sensorimotor Account of Vision and Visual Consciousness«, in: Behavioral and Brain Sciences 24/5, pp. 939–973.

Robertson, Adi (2017): »VR Was Sold as an ›Empathy Machine‹—But Some Artists Are Getting Sick of It«, in: theverge.com (05.2017). Online: https://www.theverge.com/2017/5/3/15524404/tribeca-film-festival-2017-vr-empathy-machine-backlash (last access: 03.10.2025).

Slater, Mel (2009): »Place Illusion and Plausibility Can Lead to Realistic Behaviour in Immersive Virtual Environments«, in: Philosophical Transactions of the Royal Society B: Biological Sciences 364/1535, pp. 3549–3557.

Slater, Mel et al. (2022): »A Separate Reality: An Update on Place Illusion and Plausibility in Virtual Reality«, in: Frontiers in Virtual Reality 3/914392.

Sontag, Susan (2003): Regarding the Pain of Others, New York: Farrar, Straus and Giroux.

The Protectors (2017): »Walk in the Ranger's Shoes«, in: www.walkinginrangersshoes.com. Online: https://www.walkinginrangersshoes.com (last access: 13.10.2025).

Wollheim, Richard (2015): »Seeing-As, Seeing-In, and Pictorial Representation«, in: Id.: Art and Its Objects, Cambridge: Cambridge University Press.

Yildirim, Caglar/Sercan Şengün/Mehmet Akhoroz/Eyup Kucuk/Fox Harrell (2023): »Through the Looking Glass: The Role of Virtual Mirrors in Shaping Empathy in Virtual Reality Perspective Taking«, in: Proceedings of the 22nd International Conference on Mobile and Ubiquitous Multimedia, pp. 501–504. DOI: https://doi.org/10.1145/3626705.3631789

Zucconi, Francesco (2018): Displacing Caravaggio. Art, Media, and Humanitarian Visual Culture, Cham: Palgrave Macmillan.

Image Credits

fig. 1: © Michał Stankiewicz
fig. 2–5: © John Hsu, Marco Lococo.

Orchestrating Expectations, or Technoromanticism in VR Art

Manischa Partowi

When I encounter VR in art exhibitions, I always notice some kind of hesitation. Somehow, the situation of zoning out of the shared space via this awkward tool of the head-mounted display (HMD) as a mode of experiencing art doesn't seem very promising. People in- and outside the headset act goofy, rarely seem confident or even cool. This peculiar situation seems to require a genuine neophilia and playfulness, otherwise it's loaded with social unease, shyness or embarrassment.

What is happening inside the headset that cannot share the space with me, my fellow visitors and other artworks? Do I really want to enter and potentially become affected by a space beyond the conventions of physical and institutional norms? And is my commitment big enough to involve myself in an unfamiliar situation and find out? Most of the time, it was. After being overwhelmed, excited, annoyed and disappointed by various VR experiences in contemporary art, this hesitation turned into curiosity: What is it about VR art that causes such ambivalent reactions and apparently makes me care? Of course, there are various topics and artistic styles that configure each single experience, and certainly my own taste and mood play a role in this as well. But in this text, I'm interested in a general component of the aesthetics of VR art that relates to the intensity of its experience and the concurrent exposure of the VR viewer – the exposure to the virtual space as well as to the exhibition space. Departing from the exhibition space, I want to characterize this hybrid constellation of VR art as an orchestration[1] of expectations, which navigates between attachment and detachment to both of these realms.

1 Thanks to Christian Grüny who suggested the term ›orchestration‹ as a fitting English equivalent for the form of ›Inszenierung‹ I'm discussing in this text.

I will focus on two artworks that explicitly deal with this double exposure of the viewer by creating elaborate spatial installations as a stage and as an onboarding for the virtual experience. These installations highlight the peculiar choreography of a VR experience in an exhibition space. Moreover, they communicate bodily involvement into the virtual situation, based on a form of virtual representation of the viewer. They entertain the fantasy of a technologically induced unity between body and virtual space and thus a detachment from the exhibition space. As this mystifies the VR experience and promotes the figure of an awe-driven viewer, indulging in the immersive effects of the technology, I argue, both artworks employ a technoromantic imaginary of VR as an aesthetic tool to orchestrate the expectations as well as the experience itself. While the first example tends to reinforce this imaginary of VR, the second example exposes its underlying premises.

1. Prelude

On first sight, my examples could not be more different to each other: Jordan Wolfson's *Little Room* (2025) consists of a complex technological setup, in which the viewer's body is being scanned and transferred into the virtual space to serve as an avatar. It provides a virtual experience for two people who share the same space within the body of the other. The announcement of the artwork promises a »complex intersection of real, virtual, and imaginary realms.« An artwork that »examines the darker aspects of the human experience while raising profound existential questions about consciousness, identity, and physical and intellectual negation. With *Little Room* the artist pushes the potential of VR to create a unique encounter that goes beyond conventional understandings of the medium.«[2]

Theo Triantafyllidis' *Staphyloculus (or, the paradox of site specificity of virtual realities)* (2017) is a single user experience placed in the fictional narrative of a virus outbreak in the desert. The announcement describes it as a »Virtual Reality experience that attempts to recreate the first known outbreak of Polywobbly Fervenitis [...] The artist attempted to setup a complete VR rig under

2 Description of the artwork on the homepage of Fondation Beyeler where the artwork premiered from 01.06.-03.08.2025: Fondation Beyeler (2025): »Jordan Wolfson: *Little Room*«, in: fondationbeyeler.ch. Online: https://www.fondationbeyeler.ch/ausstellungen/vergangene-ausstellungen/jordan-wolfson-little-room (last access: 15.09.2025).

scorching sunlight. [...] when the headset was exposed to the desert sun, its display had a chemical reaction that was somehow translated into a digital signal that started a computer virus.«[3] The technology as well as the artist are part of this narrative, which also »goes beyond conventional understandings of the medium« but projects a playful character onto the virtual experience. These descriptions of the artworks not only announce various forms of bodily involvement into the virtual situation but also represent different artistic gestures that already inform the anticipation of two diverse viewing experiences. *Staphyloculus* suggests a narrative-driven, experimental and rather gamified VR situation, of which the artist is part. *Little Room*, however, presents itself as a serious approach to existential questions of virtual embodiment and reinforces its conceptual gravity by highlighting the artist's objective. What both artworks address is the invasiveness of the VR technology, affecting the viewer's body as a means of experiencing the virtual space. In *Little Room*, the technological invasiveness appears in a literal sense through the body scan and the body swap. In *Staphyloculus* the technological invasiveness appears in a figurative sense in form of the virus outbreak infecting the viewer's body through the VR hardware.

Both artworks project extreme scenarios, in which the viewer's body is subjugated to the technology. As such, they address VR as »the most radical form of insertion of a user into a virtual environment.«[4] They figure the virtual experience as a commitment to an exceptional experience, beyond the familiarity of the everyday and the conventions of the exhibition space. This expectation is not only being triggered by the announcement of the artworks but further orchestrated by how the artworks are installed within the exhibition space. The spatial installations serve as a stage as well as an onboarding for the VR experiences. As such, they constitute an inside and an outside audience and explicitly address the hybrid constellation of VR art. In this constellation, *aboutness*[5] and *immediacy*[6] collide. Exposed to the outside audience, the VR experience is subjected to speculation: What kind of experience does the visceral immediacy, the

3 Description of the artwork on Theo Triantafyllidis' homepage: Theo Triantadyllidis (2017): »Staphyloculus«, in: slimetech.org. Online: https://slimetech.org/works/staphyloculus (last access: 15.09.2025).

4 Christiane Paul (2023): Digital Art, London: Thames and Hudson, p. 114.

5 Arthur C. Danto (1981): The transfiguration of the commonplace: a philosophy of art, Cambridge, MA: Harvard University Press.

6 Anna Kornbluh (2024): Immediacy. Or, the style of too late capitalism, London: Verso Books.

bodily affection by the technology produce? What is this about and how is it being expressed by the bodily behavior of the VR viewer? Exposed to the virtual situation, the viewer navigates between immediate affection by this new environment and the expectations from outside, searching for a symbolic »crack in the smoothness of the imaginary.«[7] In this hybrid situation, the ›outside‹ experience of the artwork regards a spatial as well as a temporal outside of VR. Before and after the ›inside experience‹, the viewer is part of this outside audience, which is itself co-present during the VR experience. Usually, this exposure to the »shared social norms of the actual space [...] remain[s] a constraint upon actions in the virtual space«[8] but in these examples, the particular orchestration of the viewing experience creates a situation for experiment and play.

Both of my examples deal with the egocentric dimension of VR, with what Hito Steyerl calls an »aesthetic of isolation.«[9] She describes the VR experience as a form of »bubble vision« which is subjected to a space designed to be convincing. She observes a paradoxical effect, which is the phenomenon of being at the center of the virtual world but mostly missing the actual body. My examples address this bodily paradox by experimenting with different forms of the viewer's representation within VR. From the outside, they do so by mystifying the embodied VR experience, as they propose a particular devotion to the machine to become bodily represented in VR. They trigger the expectation of a convergence between human body and the technology, which echoes Steyerl's claim: »VR is where the nonexistent bumps into the

7 For an elaborate discussion on the relation between the imaginary of immediacy and the symbolic see: Ibid., p. 62.

8 Eryn Parker/Michael Saker (2020): »Art Museums and the incorporation of virtual reality: Examining the impact of VR on spatial and social norms«, in: The international Journal of Research into New Media Technologies 26/5-6, pp. 1159–1173, here: p. 1168, https://doi.org/10.1177/1354856519897251. This form of VR-exposure shows crucial parallels to the outside parameters that prevent affective viewer's commitment in formats like one-on-one performances. See Rachel Gomme (2015): »Not-so-close encounters. Searching for intimacy in one-to-one performance«, in: Participations. Journal for Audience & Reception Studies 12/1 (01.05.2015).

9 Hito Steyerl (2018): »Bubble Vision. Aesthetics of Isolation«, in: Penny Stamps Lecture, Penny W. Stamps School of Art & Design on www.youtube.com (25.01.2018). Online: https://www.youtube.com/watch?v=T1Qhyo_PCjs&list=PLIa7Lly2iiBC3G3YCQkFE3-cOYDh5AnA-&index=4 (last access: 15.09.25).

ubiquitous.«[10] Both artworks provoke the »occult pretension [...] of VR«[11] and thus nourish its technoromantic imaginary. Technoromanticism describes an effect of individual insecurity in face of an increasingly incomprehensible complexity of technology. It frames how »mediation (whether VR or otherwise) can ›trick‹ consciousness into perceiving one thing or another – manipulating autonomic bodily processes that exceed the capacity for contemplative reflection altogether.«[12] Promoting a retreat into a position of solitude and awe, a technoromantic imaginary of virtual embodiment comes with the expectation of an immersive unity.[13] It results in a sublimation of technology in which the individual surrenders to its fetishization.[14] By diving into my own experience with both of these artworks, I want to examine how the technoromantic fantasy of VR figures within the aesthetic experience. As a tool in orchestrating my expectations from outside and as an effect of exposing these expectations from inside VR.

2. Jordan Wolfson: *Little Room*

The exhibition space of Jordan Wolfson's *Little Room* looked like a tech lab: there were lots of different stations with computers, displays, cameras and HMDs, building a complex network – not necessarily comprehensible for the regular PC user. Each station was run by several people taking care of the technology as well as of the audience. From the perspective of a visitor ready to experience VR, the room was structured by four separate VR stations each with two HMDs hanging from a scaffold at the ceiling. In the middle of the room, there

10 Ibid.

11 Richard Coyne (1999): Technoromanticism. Digital Narrative, Holism, and the Romance of the Real, Cambridge, MA: MIT Press, p. 40.

12 Marcus Carter/Ben Egliston (2024): Fantasies of Virtual Reality. Untangling Fiction, Fact, and Threat, Cambridge, MA/London: MIT Press, p. 55.

13 Ibid., pp. 19–20.

14 Kat Kitay proposes the concept of technoromanticism to grasp a trend of contemporary art after »post-internet«. In relation to Richard Coynes literary analysis, she observes an increasing »aesthetic flirtation with the materiality of technology« which, on the one hand asks existential questions about the relation between human and machine and at the same time risks to »blur[...] the lines between critique and commodity fetishization.« Kat Kitay (2024): »What's after Post-Internet Art?«, in: spikeartmagazine.com (28.02.2024). Online: https://spikeartmagazine.com/articles/essay-whats-after-post-internet-art (last access: 15.09.25).

was a station where multiple cameras (96) were mounted on a dome-like scaffold. They were directed towards the middle to capture a person's body from all around (fig.1).

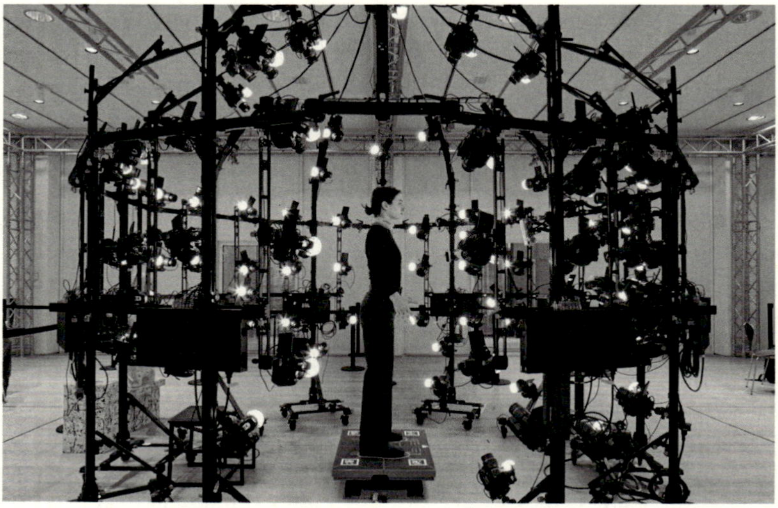

fig. 1: Jordan Wolfson: Little Room, installation view: Fondation Beyeler, Riehen, 01.06.-03.08.2025

Upon entering the exhibition space, the audience's attention was directed to this station. Guided by a retractable queuing system, we would be kept away from the HMD-settings and stand in line watching other people's bodies getting captured and waiting to go through this procedure ourselves. Someone from the staff asked me if I had a partner with whom I'd do the experience together. As I came without company, they wanted to match me with someone. After waiting a while, I did so myself, as the person seemed too busy handling other stuff like informing impatient visitors that the whole procedure would take up to one hour or preparing the next couple for the body scan.

Randomly, I matched with a Mexican gallerist who was there for Art Basel and who never experienced VR before. Knowing that we would swap each other's bodies in VR, I attempted to establish some familiarity between us. For me, this was an interesting match, as we had quite contrary positions in terms of experiences and expectations towards the artwork. Their motivation to experience the artwork was driven by curiosity to finally experience VR and by the mystery surrounding this artwork. For them, the mystery was created by the speculative formulation within the announcement of the artwork as well as by friends who had recommended the artwork without revealing much about their experience. Well aware of Wolfson's fascination with *body horror*,[15] I shared some conceptual considerations I had on a previous experience with a VR artwork of his. While this was intended to unravel some of its mystery by exposing my expectations towards some kind of disturbing, yet conceptually approachable experience, I realized that my speculation just added another level of mystification to the artwork.

We had to interrupt our small talk to prepare for the body capture. Like in a security check of an airport, we had to leave our personal belongings with the staff outside the scanner. I also had to tie back my hair and tuck in my skirt to create clear body contours. One after another, we stepped into the dome. A person who was running the technology asked us to position ourselves in a relaxed A-pose and stand still until the flash of the cameras signaled that our image would be taken. After leaving our names with this person, we were guided by another helper to a table with two chairs. On a separate sheet we had to fill in the color of our eyes, which then was taken to two people, sitting behind big monitors connected to gaming computer. The monitors were turned away from us and the people behind them were facing the whole scenery in the middle of the space – that's where the magic happens, I thought. Here the photogrammetry software would reassemble our bodies, fragmented into 96 images, to become our digital avatars.[16]

Sitting at the table, we were now positioned close to one of the HMD-settings. We observed two other people being guided into the headsets and through a seemingly complicated calibration process conducted by yet another person. This person was running in between the HMD-people and the computer station, checking the functionality of the interface, becoming increasingly exasperated by the visitors who apparently had a hard time following their orders to bodily calibrate the headsets. For my partner and me,

15 See for example: Grant Bollmer/Katherine Guinness (2020): »Empathy and Nausea: Virtual Reality and Jordan Wolfson's Real Violence«, in: Journal of Visual Culture 19/1, pp. 28–46, https://doi.org/10.1177/1470412920906261.

16 Since light-based scanners have difficulties scanning reflecting surfaces, our eyes would be added to the avatars afterwards.

this whole situation became increasingly absurd. The technical effort that was being performed around us kicked us out of an initially flowy small talk. Waiting another twenty minutes in our seats, I tried to pick up the talk again by mentioning this gigantic setup. Intimidated by its apparent complexity, we focused on the performance of the viewers who finally immersed in a virtual experience in front of us. They were goofing around with their bodies, looking at each other from various perspectives and sometimes touching each other – shaking hands or hugging. Other couples behaved differently, only looking at each other or performing careful movements with their bodies. My partner admitted that they were not sure whether they even wanted the experience anymore. They exposed their insecurity about the whole situation. The longer we waited, the closer we got to the virtual experience, initial genuine curiosity turned into uncomfortable indisposition. How did we have to behave within the virtual? How close would we get? And how would it feel inhabiting the other's body? I was also uncomfortable with this projection but unlike my partner, really looking forward to it – given this hypercomplex technical and personal orchestration of the experience, I expected a lot. At least something that would be an answer to this spectacle.

We entered the VR by standing next to each other, each of us equipped with a headset. To calibrate, I was told to take my hands in front of my face, stretch one arm away from me, the other one close to my headset, facing the palm. Slowly walking a few steps forward, a few steps backward, I was continuously switching this position with my arms back and forth. After this was done, I put on a pair of headphones and waited for my partner to be ready. The appearance of a white, empty space with a few grid-like lines signaled the start of the virtual experience. We immediately faced each other. Instead of my partner, I saw a three-dimensional image of myself in front of me. It looked like an empty shell, like a puppet.[17] One of my arms had a strange distortion. Also, there were some glitches at the intersection of different materials, such as the skin and the cloth of my t-shirt or at my hairline. Even though the retrospectively added brown eyes were staring at me, without a blink, I found this puppet of myself quite cute. I had an amused look on my face, which made me think

17 For an extensive discussion of the relation between human and avatars see for example the contribution by Federica Cavaletti/Pietro Conte/Andrea Pinotti to this volume. See also Dieter Mersch et al. (eds.) (2023): Actor & Avatar. A Scientific and Artistic Catalog, Bielefeld: transcript.

back to the situation in which this image was taken. Also, the very visible technical errors were charming, as they highlighted the technical artificiality of this bodily object. But my partner had a different impression of themselves. They didn't want to look at me/themselves, expressing that they didn't know that they were so ugly. Bewildered by this reaction of my partner, I tried to convince them that it was the digital capture making us look like this: »It's their errors, not ours. Look, our mouths do not move, even though we are talking to each other.« But this did not work: »I'm so old and look at my posture!« I looked down at my actual body and observed the virtual representation of my partner. Their visual appearance was the same as mine, only as them. I looked at my/their hand and turned it around. Moving this body felt like a typical avatar representation in VR. The movements of the body parts that were not tracked – which included everything else than the hands and the head – were stiffly and schematically simulated. After a few moments, a mirror flew into the virtual space. We positioned each other in front of it and looked at our virtual bodies. Now I could guess why my partner was disturbed by their representation: their facial impression was rather shocked, which indeed looked a bit scary with the staring eyes. I took the mirror as a chance to project my own liveness into the avatar. I/they waved into the mirror, using my avatar as a tool to signal my sympathy but I wasn't sure about delivering this message. Neither could I attune to my avatar as a body, nor could we establish any relation to each other with or beyond these empty shells.[18] The mirror disappeared and suddenly we heard the voice of the artist through the headphones. The mouth of my partner/my mouth opened abruptly: »I am your father.« A different voice: »I'm your mother.« Him again: »God molested you.« The other one: »I'm God.« ... »God murdered me.« ... Statements like these were exchanged back and forth. I was assuming that my partner experienced the same as I did, seeing the mouth of my avatar opening and closing mechanically, simultaneously to the audio overlay. The last sentence was »Look at your hands.« Doing so, the virtual space vanished around us. We took off the hardware and looked at each other. My first thought was: »Seriously?« – which I probably also expressed on my face. Curious about the impressions of my partner, I asked them if we should take a minute and talk about it. »No, this was strange,« was the answer, and they left immediately.

18 For the concept of attunement between body and avatar see the contribution by Ksenia Fedorova/Jose Hopkins B. to this volume.

After spending more than an hour in the exhibition space and only about twelve minutes within the VR, I couldn't leave directly. I sat down on a chair to digest – not an existential disturbance but a profound irritation about the unimpressiveness of my VR experience. Apparently, the voice-over was supposed to add psychotic heaviness to the exerted body swap. But in all its artificial constructedness, it appeared to me as a lesson on what I was supposed to feel – intruded and possessed by the other. I was annoyed by this godlike performance of the artist, of which the voice-over was only the peak. The whole orchestration of the situation – the commitment of the visitors to wait extremely long to mess with their bodily integrity, the endeavor of 20 people in service of the technology and of the anxious, impatient, disturbed or also just disappointed audience – all of this seemed quite off.

But what was my partner disturbed and I disappointed of? Apparently, the whole situation has caused strong and ambivalent reactions that I think are revealing of the technoromantic myth that sustains this artwork. At the center of this seems to be the strange form of embodiment and the phenomenon of becoming affected by it. It's not about an affection towards the body you are inhabiting but towards the representation of your own body that appears in front of you. In various articles, this affection by the own bodily representation has been described as form of »cold mutual possession«, figuring as a »mutation of the body and a profound loss of agency«, a »dissociation of our own bodies in the most visceral way.«[19] While my partners' immediate reaction might fit to these descriptions, I could not establish any kind of visceral relation to my bodily representation. I was indeed disappointed that I did not experience any of the expected intensity and weirdness of embodiment. Instead, I felt affected by its technical curiosity. It was this »irritating hybrid of human and machine, of flesh and cyberspace«[20] that represented my image-body as a puppet and prevented any physical form of possession or dissociated embodiment. Rather than detaching me from the technological infrastructure, the simple arrangement of our clumsy bodies in this white, gridded space reminded me of the

---

19 Kate Brown (2025): »Jordan Wolfson's New Virtual Reality Is Peak Body Horror«, in: artnet.com. (24.06.2025). Online: https://news.artnet.com/art-world/jordan-wolfson-little-room-2659690 (last access: 15.09.2025).

20 Jolinde Hüchtker (2025): »Komm, wir tauschen unsere Körper. Eine Installation des Künstlers Jordan Wolfson macht's möglich«, in: zeit.de (13.07.2025). Online: https://www.zeit.de/2025/29/jordan-wolfson-koerper-tauschen-little-room/seite-2 (last access: 15.09.2025).

interface of a game-engine. Instead of nourishing a sense of psychotic dissociation, this »clinical coldness«[21] of the virtual space placed me into some kind of experimental stage of VR, maybe in a context of play-testing the avatars. While this could have provided access to a more experiential reading of the artwork's technological conditions, the voice-over constituted the actual source of disappointment. It presented me some kind of reading of the artwork, which I couldn't relate to myself – the disturbing dimensions of missing attunement between physical body, body image and a sense of the self. Within the framework of play-testing, the voice-over appears as yet another layer of the experiment, exposing the immediate bodily affectedness as devotion to the illusional character of the technology.

Apparently, there are very different readings of this artwork which relate to the experiences and imaginary capacity of each viewer. However, I argue that the celebration of body horror in this artwork relates to a particular detachment from the exhibition space. It requires ignoring the technical and social efforts around the VR experience and presupposes the belief in its power to produce an immersive effect of unity between body and machine in order to be effectively disturbed. The egocentrism of immersive experience is being reproduced and backed up by the psychoanalytical framework of an »oedipal drama triangle between father, mother, and child, with alternating victim and perpetrator roles«[22] is being introduced by the voice-over of the artist. This reading features the mystification of the technology, as it confirms VR's immediacy trick as a mode for rehearsing the psychotic dissonance of individual's self-perception. Linking this experience back to the concept of technoromanticism, the exposure of the technological infrastructure and the laborious procedure of bodily involvement in VR could have led to disenchanting the technology.[23] Yet, the actual procession of the images, the work of the software, stays hidden inside the networked hardware, awaiting the viewer inside the headset for subjugation. The way the audience is directed towards the VR spectacle not only fetishizes the technology but also inflates the expectations towards the

---

21 Sebastian Frenzel (2025): »VR-Installation von Jordan Wolfson: Ist man Gott begegnet, ein wenig gestorben und wieder geboren?«, in: monopol-magazin.de (29.06.2025). Online: https://www.monopol-magazin.de/jordan-wolfson-vr-installation-little-room-fondation-beyeler (last access: 15.09.2025).

22 Ibid.

23 For a discussion of the semiotics of enchantment of VR technology see Stefan Rieger (2024): Imagination und Immersion. Topologien des Virtuellen, Bielefeld: transcript, esp. pp. 323–324 and p. 402.

virtual experience, which tends to figure as a self-fulfilling prophecy.[24] Aiming at a detachment from technological infrastructure in favor of its effectiveness, I argue that the virtual situation promotes an awe-driven viewer, enforcing the technoromantic myth rather than disclosing it. Rehearsing and exposing the individual's helplessness in face of the technology's effectiveness, the artwork tends to »entertain[...] gloom-and-doom pastiche without saying anything at all.«[25]

What this »aesthetic flirtation with the materiality of technology«[26] does show is the enormous infrastructural effort to produce and provide a more or less convincing virtual viewing experience. It also exposes a crucial phenomenon of the hybrid constellation of VR art, namely the social awkwardness in face of the anticipated bodily affection by the virtual situation, arising from the orchestration of expectations within the exhibition space. The explicit exposure of the technology, the aesthetic flirtation with its materiality, causes and at the same time highlights the fundamentally unsettling effect of VR art and the genuine insecurity in how to properly approach the artwork. This unsettling effect of the technology is backed up by the narrative of the artwork. Spreading within and beyond the walls of the exhibition space, even into the virtual space of the headset, this narrative demonstrates the power of »rumor as a participatory force«[27] as a crucial element in the artistic orchestration of expectations.

3. Theo Triantafyllidis: *Staphyloculus (or, the paradox of site specificity of virtual realities)*

While also staging the awkwardness between inside and outside audiences, my second example turns the effects of a technoromantic imaginary around. It hides the technological complexity of virtual embodiment behind the expectation of a playful, even random experience. In *Little Room*, playfulness was the

---

24 See Richard Jenkins (2000): »Disenchantment, Enchantment and Re-Enchantment: Max Weber at the Millennium«, in: Max Weber Studies 1/1, pp. 11–32, esp. p. 13.

25 K. Kitay: What's after post-internet Art?

26 Ibid.

27 Karin Bruns (2006): »›Do it wherever you want it but do it!‹ Das Gerücht als produktive Partizipationskraft der ›Neuen Medien‹«, in: Britta Neitzel/Rolf F. Nohr (eds.), Das Spiel mit dem Medium. Partizipation, Immersion, Interaktion, Marburg: Schüren Verlag, pp. 332–347.

least expected mode of aesthetic experience, in *Staphyloculus* it is the most expected one, which turns out to be just right for exploring the profoundness of the artwork.

The exhibition setup consists of a simple arrangement of VR-hardware, a banner, and a second screen. The VR viewer, equipped with the bulky headset of the HTC-Vive and its two controllers, is standing in front of a horizontal banner, which is suspended from the ceiling (fig. 2). It shows a panorama view of a desert. Various images are collaged onto this backdrop, creating a peculiar scene: There is a VR-setup in the middle of the desert, surrounded by a few big rocks. The headset and the controllers are floating in between two room-scale sensors. A red location pin is set in the distance. Around this setting, some coral-like creatures lurk in the vegetation. There is a big drawing on the right side of the banner that looks like a depiction of the biological life cycle of these creatures. It connects the following pictograms with each other: a few differently shaped creatures – human eyes looking in different directions – one human eye labeled with scientific terms – two hands holding a gamepad – a schematically depicted figure in T-pose – and something that looks like a skin infection. One of these coral-like creatures appears as a physical model, being presented by a human hand from behind of a rock on the left side of the banner. In front of this rock, a monitor and a keyboard are placed on the desert ground. Close to this setting, a strange, small sphere is hovering above the ground. It shows a person with an HMD on their head, standing in the same desert location as that of the panorama. The person is directed towards the rock – the one in the sphere as well as the one in the panorama. Following the perspective of the person to very left edge of the banner, Caspar David Friedrich's wanderer appears on a rock at the horizon, gazing into the supposed vastness of the desert beyond our point of view. Our point of view on this setting and on the VR viewer in front of the banner is also being emphasized by a shadow of a figure cast onto another rock.

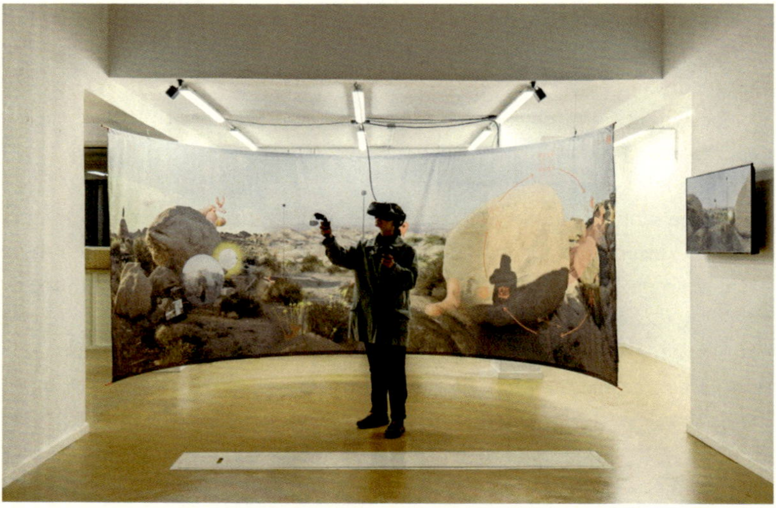

fig. 2: Theo Triantafyllidis: Staphyloculus, 2017, installation view: Plural Landscapes, group show at Keiv, Athens, 01.-17.04.2022

But what are we seeing from this point of view? How do the different images of the desert, the creatures and the technology relate to each other and to the virtual experience? The collage-like arrangement on the banner resembles some kind of storyboard for the VR experience, but without a comprehensible narrative. As a backdrop for VR experience, it rather seems like a mood board, visually placing the viewer in a unique atmosphere. This setting explicitly exposes the behavior of the VR viewer to the exhibition space which turns into a performance ready to be interpreted by an outside audience. But this performance is similarly incomprehensible. The VR viewer gestures with the controllers and maneuvers their body in peculiar ways. Sometimes, it looks like they are eagerly interacting with something stubborn in digital space. Sometimes they just seem to observe their surroundings. Competition or contemplation? Intuitive play or choreographed movement? The screen, which is mounted onto the wall next to the VR-setting, could provide information about the actual experience. But instead of translating the situation from the point of view of the immersant, as the second screen in a VR setup usually does, it depicts yet another perspective onto the scenery that must be captured from inside the virtual scenario. Here, we can see the VR hardware moving in the desert according to the viewer's position. In this sense, the viewer's bodily movements become integrated into the desert scene, but what the actual experience consists of stays mysterious. The wall label next to the screen is giving a hint:

> »A Virtual Reality experience that attempts to recreate the first known outbreak of Polywobbly Fervenitis. According to live footage found on the site in Joshua Tree, California, the depicted artist attempted to setup a complete VR rig under scorching sunlight. As suggested by the investigation, when the headset was exposed to the desert sun, its display had a chemical reaction that was somehow translated into a digital signal that started a computer virus. The computer virus that had some kind of primal Artificial Intelligence, managed to mutate itself into a strain of bacteria, now known as Staphylocu-lus, that spreads through the use of Virtual Reality headsets. In this first incident, the artist was infected by the microorganism and, after hosting it for a while, brought it back to Los Angeles and started the spread of Polywobbly Fervenitis. The ability of this sentient microorganism to freely transform from computer virus to real world bacteria has astounded the scientific community that is anxiously looking for a cure.«[28]

This is providing information about the images on the banner: the location must be in the national park, the VR setup refers to the technical experiment, the image of the sphere probably documents the artists conducting it and the coral-creatures are likely to be the virus whose outbreak is depicted in the life cycle. Yet, the situation of the viewer stays blurry: Is this a recreation of the outbreak in which we're going to be infected? Or are we part of the scientific community that is looking for a cure? This description of the artwork creates a backstory to the virtual experience and situates the viewer within a science-fiction story. Instead of giving reasonable background information on the artwork, it figures as a lore, known from computer games, which creates a narrative to prepare the user for its role within the virtual environment. As such, the VR experience is being detached from the conventions of the exhibition space and suggests a playful mode of experiment and play.

Knowing that my actual experience would not be translated onto the screen, which prevented the outside audience from being able to judge my behavior, I felt encouraged to go for this playful attachment to the virtual space. Inside the VR, I landed in the desert. The vastness of the desert appeared like

28 Website Theo Triantafyllidis.

a scene from an open world game, ready to be explored.[29] Intuitively, I tried to use the controllers to teleport myself through the space. As this did not work, I was restricted to explore my immediate surroundings. The carefully designed visual and acoustic details – the warm light of the sunset, the shadows of the vegetation, and the close humming of an insect – addressed a sense of tactility. My body itself was missing, it was only represented by the headset and the controllers, which cast a shadow on the desert ground. I could also observe this body representation within a monitor, which was placed in between the bushes and served as some kind of mirror. This monitor was part of a complete VR rig that was surrounding me in virtual space, similar to the technical setup in the exhibition space as well as in the artist's experiment.

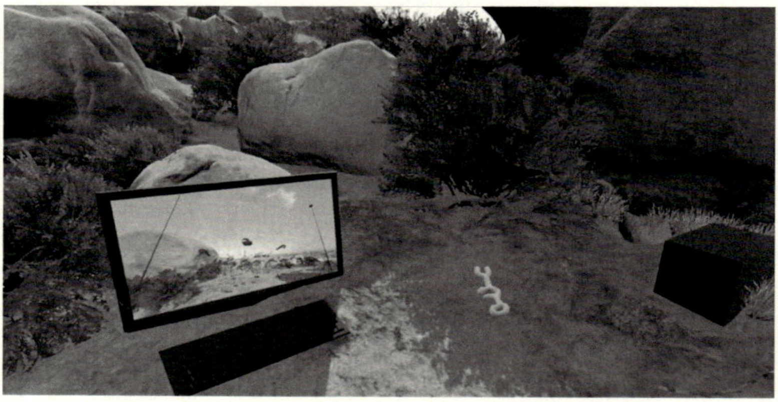

fig. 3: Theo Triantafyllidis: Staphyloculus, 2017, ingame screenshot from the VR-experience

Suddenly, some of the small creatures appeared in the vegetation, shyly hiding from my gaze behind the bushes and rocks. They drew my attention away from the technical setup and one after another jumped towards me with a cute, squeaking sound. When I tried to grab them, they attached themselves

29 Marc Bonner (2018): »Die gekerbte Wildnis – Inszenierungen vermeintlich unberührter Umwelt in digitalen Spielwelten«, in: Paidia. Zeitschrift für Computerspielforschung (28.02.2018), pp. 1–24.

onto my hardware. Like a skeleton, they bridged the gap between headset and controllers. On one hand, I did see the urge of preventing this bodily invasion by trying to get rid of these stubborn creatures, on the other hand, this arbitrary visual representation of my body added a feeling of pleasure to my embodiment. I played around with this body and observed myself in the shadow and in the monitor (fig. 3). I used the technology to observe my whole virtual body– how it would bend down to the screen and move back again. Apparently, this technical mirror supported my sense of attunement to the virtual body, which I playfully explored.

After a few minutes, the wobbly creatures burst with a plop and my bodily representation disappeared. Instead, a video sphere appeared in front of my face, mirroring the whole scene in which I was standing (fig. 4). The artist appeared close to the surface, pulling his headset over his eyes. As he started to explore his surroundings with the controllers, his voice appeared from off screen: »Ok, this is a camera... 360 camera right here... The rock, is here... aaah, this is a real rock! What is this? Where are we?... Ok, I can walk around. I can feel the wind on my face. I can feel the wind on my fucking face!« He was commenting on his experience like in a let's play video. This situation was yet another mirror of my previous experience, but not in a sense of supporting virtual embodiment but of exposing my virus-body as a fictional illusion of what I was actually doing: gesturing in space with the technology. While the virus was my tool of embodiment, the artist's tool is the technology on which he commented further. To observe whether our current situations would also relate to each other, I tried to touch the same rock as he did, but in my surroundings. Instead of feeling anything, a coral-colored guardian appeared in front of my face, marking the technical boundary of my freedom of movement as well as of the lacking materiality of my virtual surrounding. I turned towards the rock in the video, which inflated as soon as I touched it with my controller. An animated sound corresponded to this inflation and also the voice-over reacted to it: »aaah, its growing, its growing... So, this is a skybox? Or is it a video?« The video sphere inflated until it encapsuled me (like a skybox). Suddenly, a wireframe stretched over the whole scenery, covering the surface of the desert with a coral-colored polygon grid. The artist next to me tripped over a cable: »What am I supposed to get out of this? Come on – virtual reality? You call this virtual reality? What the fuck is this?« The virtual surrounding burst into multiple particles and ended in a coral-colored void.

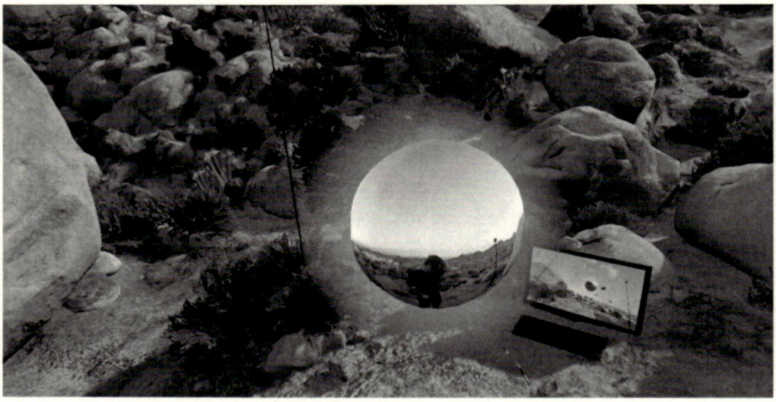

fig. 4: Theo Triantafyllidis: Staphyloculus, 2017, ingame screenshot from the VR-experience

Indeed, what was this? How do both parts of this artwork, the peculiar embodiment through the technology as well as the virus and the eversion of the virtual space, relate to each other? The appearance of the video sphere served as some kind of reality check. Seeing the artist handling the technology and hearing his comments, I realized that I had not thought about how my situation was constructed. Instead, I fully immersed myself into the playful affection by the virus. Like the virus, the representation of the technology functioned as a tool for further bodily involvement. Reconstructing the virtual scenario as a bubble and revealing my genuine bodily involvement as a convincing effect of its egocentric and affective tools,[30] the video sphere destabilized this experience of unity between my body and the technology. It exposed the fragility of immersion into a virtual situation and unraveled all kinds of layers of its constructedness. By addressing the technological components constituting the visual space, such as the 360° camera, a wireframe, the video and a skybox, the experienced stability of the environment becomes questionable: The wireframe exposes the desert environment as a scan. The correlation between 360° video and a skybox links back to the infrastructure of a game-engine where all kinds of visual material – be it the scan or the video – may serve as a resource for spatial simulation and, in case of the skybox, for the illusion of

30 H. Steyerl: Bubble Vision.

infinity.[31] The video sphere links the question of space to that of body and proposes an understanding of one's own embodiment as part of the artists experiment. From this perspective, the virus body emerging in between the VR-hardware can be addressed as an exploration of the VRIK model (fig. 2). This is a »full body solver« that tracks the position of controller and headset to simulate the whole body with the means of an inverse kinetics algorithm.[32] The experience of affective attachment, the effect of bodily attunement with the skeleton of this body model highlights the paradoxical effect of »double bind patterning,« which furnishes the inanimate digital object with a »quality of liveness through movement.«[33]

This conceptual link between both parts of the artwork is supported by the narrative of the virus outbreak. At first, it's attached to the viewer's body, then it's attached to the environment figuring as a mutation in form of the coral-colored wireframe and the guardian. Just like the virus constituted the skeleton of my virtual body, the wireframe constitutes the skeleton of the virtual environment, revealing the perceptual illusion of materiality and liveness as an experienced relation between skeleton and skin. The exploration of the artist and his let's-play like commentary provide access to these technical layers of observing the phenomenon of embodiment within the virtual space. At first, I felt exposed by the disclosure of my lived illusion. But against this background, the playful affection by the virus outbreak seems to be an important detour to this observation: to know what you're observing you need to experience yourself.

Coming back to the expectations orchestrated by the exhibition set-up, I understand the inflated narrative of the science-fiction scenario as a tool to detach the VR experience from the exhibition space. Directed to an outside

31 Marc Bonner (2021): »The World-Shaped Hall. On the Architectonics of the Open World Skybox and the Ideological Implications of the Open World Chronotope«, in: Id. (ed.), Game, World, Architectonics. Transdisciplinary Approaches on Structures and Mechanics, Levels and Spaces, Aesthetics and Perception, Heidelberg: University Publishing, pp. 65–98.

32 Big thanks to Theo for taking the time for a studio visit and walking me through the technical layers of the artwork. Technical introduction of VRIK on the oculus developers page: https://developers.meta.com/horizon/blog/developer-perspectives-character-animation-in-dead-and-buried/ (last access: 15.09.2025).

33 Sina Seifee (2022): »Rigging Deamons«, in: Blanca Pujalset et al. (eds.), DATA browser 08 Volumetric Regimes. Material cultures of quantified presence, Open Humanities Press, pp. 32–56, here: p. 39.

audience, it delivers a gamey backstory as a framework for interpreting the viewer's performance in relation to the collage-like banner and the perspective of the second screen. The fact that this setup doesn't expose the viewer's actual experience supports a detachment from the exhibition space and an adaption of a curious and playful mode within the virtual space. The role of the viewer as someone becoming infected and looking for a cure translates into an affective as well as reflective negotiation of the technological and imaginary effects of inhabiting virtual space with a virtual body. In the virtual experience of the artwork, the initial game world situation is irritated by a form of 360° documentary of this situation, which is infected by the narrative itself. It is not a question of a distinct or even transcendent outside perspective on the VR experience but the interrelation between different modes of reception, which addresses the »paradox of site specificity of VR« as the central element of its aesthetic experience. Initially, the virus narrative tends to cover the complex interplay between body, space and technology. In a very explicit sense, this work description employs rumor as a participatory force to involve the viewer into some kind of fantasy. Taken as a gamey backstory, this science-fiction scenario risks to »diminish tangible concerns with equipment and embodiment.«[34] Integrated into the actual experience of the artwork, this narrative is all about these tangible concerns with equipment and embodiment. While projecting the fantasy of a virtual space, in which the »boundaries of human and machine will vanish,«[35] it introduces the role of imagination into the production of an elusive unity of virtual embodiment and thus complicates its material parameters of body and technology. As such, I read the artwork as a humorous commentary on the technoromantic fantasy, as a revelation of the infrastructural conditions of VR's immediacy effect and as a motivation to use of our affective fascination as the basis for active engagement with its material reality.

Conclusion

Coming back to my question from the beginning: What is it about the hybrid constellation of VR art that causes strong and ambivalent reactions? Feeling excited, annoyed, overwhelmed and disappointed – all of this was part of my

34 R. Coyne: Technoromantism, ix.

35 Ibid. p. 19.

experiences of the artworks. Apparently, they are indicators for strong affection by the artworks, which is the result of a fundamental commitment to their propositions. In the discussion of these examples, I was interested in how both artworks deal with the hybrid situation of the viewer – what is being experienced of the artwork from outside, how does this affect the experience of the inside and how do both spheres interrelate in terms of aesthetic experience? Expectations about the relation between body and technology, about its lived intensity versus its (in)comprehensible complexity are what migrates between outside and inside experience. They become visible by being disappointed. Instead of leading to a dead end, this provides further access to the artistic gesture in playing with this relation and its imaginary.

I was excited by the anticipation of *Little Room* prompted by the lure of body horror inside as well as by the technical show-off outside. Yet, both of these dimensions of the artwork were disappointing given its clichéd sensationalism. The virtual space of *Little Room* traps the individual inside its own insecurity about virtual embodiment and features a position of helplessness towards overwhelming effects of the technology.

I was not necessarily excited by the anticipation of *Staphyloculus* prompted by the imagination of becoming infected as well as of playing a game in the exhibition space. Yet, the disappointment of these expectations nourished a sense of excitement in face of partaking in the disclosure of its sensationalism. The virtual space of *Staphyloculus* plays with different modes of virtual embodiment and features the individual's insecurity as a means of dealing with the overwhelming effects of the technology.

Both of these artworks address virtual embodiment as a question of imaginary and tactile attunement to a virtual image body. They stress the fragility of this phenomenon by questioning the materiality of virtual representation. Yet, they draw crucially divergent figures of the viewer – one whose affection further promotes this fragility, and one whose affection is part of the questioning. When I put both artworks in Hito Steyerl's crystal bubble and look at the figure of the viewer from outside – in both of them, the viewer's body is still missing. It's lost in the isolation of *Little Room* and encouraged to deal with this lostness by *Staphyloculus*.

Bibliography

Bollmer, Grant/Katherine Guinness (2020): »Empathy and Nausea: Virtual Reality and Jordan Wolfson's Real Violence«, in: Journal of Visual Culture 19/1, pp. 28–46. DOI: https://doi.org/10.1177/1470412920906261

Bonner, Marc (2018): »Die gekerbte Wildnis – Inszenierungen vermeintlich unberührter Umwelt in digitalen Spielwelten«, in: Paidia. Zeitschrift für Computerspielforschung (28.02.2018).

Bonner, Marc (2021): »The World-Shaped Hall. On the Architectonics of the Open World Skybox and the Ideological Implications of the Open World Chronotope«, in: Id. (ed.), Game, World, Architectonics. Transdisciplinary Approaches on Structures and Mechanics, Levels and Spaces, Aesthetics and Perception, Heidelberg: University Publishing, pp. 65–98.

Brown, Kate (2025): »Jordan Wolfson's New Virtual Reality Is Peak Body Horror«, in: artnet.com. (24.06.2025). Online: https://news.artnet.com/art-world/jordan-wolfson-little-room-2659690 (last access: 15.09.2025).

Bruns, Karin (2006): »›Do it wherever you want it but do it!‹ Das Gerücht als produktive Partizipationskraft der ›Neuen Medien‹«. In: Britta Neitzel/Rolf F. Nohr (eds.), Das Spiel mit dem Medium. Partizipation, Immersion, Interaktion, Marburg: Schüren Verlag, pp. 332–347.

Carter, Marcus/Ben Egliston (2024): Fantasies of Virtual Reality. Untangling Fiction, Fact, and Threat, Cambridge, MA/London: MIT Press.

Coyne, Richard (1999): Technoromanticism. Digital Narrative, Holism, and the Romance of the Real, Cambridge, MA/London: MIT Press.

Danto, Arthur C. (1981): The transfiguration of the commonplace: a philosophy of art, Cambridge, MA: Harvard University Press.

Fondation Beyeler (2025): »Jordan Wolfson: Little Room«, in: fondation-beyeler.ch. Online: https://www.fondationbeyeler.ch/ausstellungen/vergangene-ausstellungen/jordan-wolfson-little-room (last access: 15.09.2025).

Frenzel, Sebastian (2025): »VR-Installation von Jordan Wolfson: Ist man Gott begegnet, ein wenig gestorben und wieder geboren?«, in: monopol-magazin.de (29.06.2025). Online: https://www.monopol-magazin.de/jordan-wolfson-vr-installation-little-room-fondation-beyeler (last access: 15.09.2025).

Gomme, Rachel (2015): »Not-so-close encounters. Searching for intimacy in one-to-one per-formance«, in: Participations. Journal for Audience & Reception Studies 12/1 (01.05.2015).

Hüchtker, Jolinde (2025): »Komm, wir tauschen unsere Körper. Eine Installation des Künstlers Jordan Wolfson macht's möglich«, in: zeit.de (13.07.2025). Online: https://www.zeit.de/2025/29/jordan-wolfson-koerper-tauschen-little-room/seite-2 (last access: 15.09.2025).

Jenkins, Richard (2000), »Disenchantment, Enchantment and Re-Enchantment: Max Weber at the Millennium«, in: Max Weber Studies 1/1, pp. 11–32.

Kitay, Kat (2024): »What's after Post-Internet Art?«, in: spikeartmagazine.com (28.02.2024). Online: https://spikeartmagazine.com/articles/essay-whats-after-post-internet-art (last access: 15.09.25).

Kornbluh, Anna (2024): Immediacy: Or, the style of too late capitalism, London: Verso Books.

Mersch, Dieter et al. (eds.) (2023): Actor & Avatar. A Scientific and Artistic Catalog, Bielefeld: transcript.

Oculus Developers Homepage: https://developers.meta.com/horizon/blog/developer-perspectives-character-animation-in-dead-and-buried/ (last access: 15.09.2025).

Parker, Eryn/Michael Saker (2020): »Art Museums and the incorporation of virtual reality: Examining the impact of VR on spatial and social norms«, in: The international Journal of Research into New Media Technologies 26/5-6, pp. 1159–1173. DOI: 10.1177/1354856519897251

Paul, Christiane (2023): Digital Art, London: Thames and Hudson.

Rieger, Stefan (2024): Imagination und Immersion. Topologien des Virtuellen. Bielefeld: transcript.

Seifee, Sina (2022): »Rigging Deamons«, in: Blanca Pujals et al. (eds.), DATA browser 08 Volumetric Regimes. Material cultures of quantified presence, Open Humanities Press, pp. 32–56.

Steyerl, Hito (2018): »Bubble Vision. Aesthetics of Isolation«, in: Penny Stamps Lecture, Penny W. Stamps School of Art & Design on www.youtube.com (25.01.2018). Online: https://www.youtube.com/watch?v=T1Qhy0_PCjs&list=PLIa7Lly2iiBC3G3YCQkFE3-cOYDh5AnA-&index=5 (last access: 15.09.25).

Theo Triantafyllidis Homepage: https://slimetech.org/works/staphyloculus (last access: 15.09.2025).

Image Credits

fig. 1: © Jordan Wolfson. Courtesy the Artist and Gagosian, Sadie Coles HQ, and David Zwirner. Photo: Mark Niedermann

fig. 2–4: Courtesy Theo Triantafyllidis

Appendix

Authors' Biographies

Federica Cavaletti (Dr.) is a Tenure Track Researcher in Aesthetics (Telematic University San Raffaele Roma, Italy) and AIAS-AUFF Fellow at the Aarhus Institute of Advanced Studies (Aarhus University, Denmark). After working extensively on VR and its usage in medicine and psychotherapy, she is currently back to studying the experience of time and how it affects our overall being in the world.

Pietro Conte is Associate Professor of Aesthetics at the University of Milan. His research focuses on illusion, hyperrealism, mimicry, and the multifarious practices of un-framing, with a particular emphasis on immersive (virtual) environments.

Annet Dekker is Associate Professor Media Studies: Archival and Information Studies, and Comparative Cultural Analysis at the University of Amsterdam and Visiting Professor and co-director of the Centre for the Study of the Networked Image at London South Bank University. Her main interests are in digital art and culture, preservation and sustainability.

Ksenia Fedorova (PhD) is Assistant Professor at Leiden University, NL. She is the author of Tactics of Interfacing: Encoding Affect in Art and Technology (MIT, 2020) and co-editor of Media: Between Magic and Technology (2014, in Russian) and special issue of Artnodes journal Theorizing Media Art in Light of STS (2025). Ksenia's interests encompass media art theory and history, aesthetics, philosophy, science and technology and visual culture studies, with a focus on the effects of technologies on human perception and interaction.

Stephan Günzel is a Professor of Media Theory at the University of Europe for Applied Sciences (UE) in Berlin since 2011. From 2018 to 2023 he has been a Visiting Professor and head of the Department for Media Studies at the TU Berlin. His research interests include spatial and image theory, digital games as media and almost any aspect related to the notion of space; these days in particular the implications of AR-, MR- and VR-media. Visit: stephan-guenzel.de.

Jose Hopkins B. is a researcher, writer, and curator based in Utrecht and Lima. He works at the intersection of new media arts, performance, and post-humanist thought, focusing on technology, storytelling, and their affective and embodied dimensions. As a PhD candidate at Leiden University, his research investigates decolonial, technofeminist, and postphenomenological perspectives on virtual reality.

Choghakate Kazarian (Dr.) is an independent curator and art historian specializing in modern and contemporary art. She previously served as curator at the Musée d'Art moderne de Paris, where she organized major exhibitions devoted to Lucio Fontana, Henry Darger, and Karel Appel. Most recently, she co-curated Immersion. The Origins, 1949–1969 at the Musée cantonal des Beaux-Arts de Lausanne, a landmark exhibition tracing the early history of immersive installations.

Manischa Partowi is a researcher interested in affective strategies in contemporary art. In her PhD project, she investigates the aesthetic constellation of virtual reality-based art by focusing on the phenomena of intimacy and violence. Trained in Art History, Philosophy and Cultural Studies, she is currently affiliated with the Collaborative Research Center »Virtual Lifeworlds« at Ruhr-University Bochum.

Andrea Pinotti (Prof. Dr.) is a Professor of Image Theory at the University of Milan and researches on visual studies, empathy theories, monuments and memorials. He is currently exploring the religious roots of contemporary digital technologies.

Chris Salter is an artist and Professor for Immersive Arts and Director of the Immersive Arts Space at the Zurich University of the Arts (ZHdK) as well as Professor Emeritus, Design and Computation Arts at Concordia University in Montreal. His artistic and research interests revolve around the development and production of real time, computationally-augmented performance environments.

Jens Schröter (Prof. Dr.) is a Professor of Media Studies at the University of Bonn since 2015. He was Professor for Multimedial Systems at the University of Siegen 2008-2015. His research interests include the computerized palate, the question of how artificial intelligence is changing science, theories and histories of digital media, as well as future media and media of the future. Visit: www.medienkulturwissenschaft-bonn.de / www.theorie-der-medien.de / www.fanhsiu-kadesch.de

Ursula Ströbele (Prof. Dr.) is Professor for art history with a special focus on contemporary art at Braunschweig University of Art. Her current research focus is on digital phenomena of the sculptural, queer ecologies, the transformation of the sculpture canon in 20th century art, infrastructural studies.

Manuel van der Veen (Dr.) is a research assistant in the sub-project *Virtual Art* within the Collaborative Research Center 1567 *Virtual Lifeworlds* at Ruhr-Universität Bochum. Previously, in 2022, he completed his doctorate on the arthistory of augmented reality. His further research interests include the philosophy of technology and space in relation to contemporary digital art, intersections between new media and painting as well as the tradition of three-dimensional image carriers.